50 NIFTY

SUPER SCIENCE FAIR PROJECTS

Compiled by Jill Smolinski

**Written by
Carol J. Amato,
Eric Ladizinsky, and Jill Smolinski**

Illustrated by Kerry Manwaring

*Reviewed and endorsed by Tiffany Anderson, M. Ed.,
Science Department Chairperson, John Adams Middle School,
Santa Monica, California*

Lowell House
Juvenile
Los Angeles

Contemporary Books
Chicago

Publisher: Jack Artenstein
Vice President, Juvenile Division: Lise Amos
Director of Publishing Services: Rena Copperman
Editorial Director: Brenda Pope-Ostrow
Senior Editor: Amy Downing
Managing Editor, Juvenile Division: Jessica Oifer
Craft Stylist: Charlene Olexiewicz
Cover Photographer: Ann Bogart

Cover craft by Anthony Bogle, sixth grader at Chaparral Middle School
in Moorpark, California, and second place winner at the
1995 Ventura County Fair, computer/math division

Library of Congress Catalog Card Number is available.

ISBN: 1-56565-364-5

10 9 8 7 6 5 4 3 2 1

Contents

Getting Started   5

    The Scientific Method   5
    What Makes a Great Science Fair Project?   6
    Follow the Rules!   7
    What's Next?   8
    Winning Touches for Your Project   9

50 Nifty Super Science Fair Projects   11

  1   The Heat Is On   11
  2   Get the Point?   12
  3   The "Eyes" Have It   14
  4   Water Weight Gain   16
  5   Are Your Suds "Duds"?   18
  6   Earn a Good Conduct "Metal"   20
  7   Hot, Hot, Hot!   22
  8   Penny for Your Thoughts   24
  9   Oh, Grow Up!   25
 10   Don't Be So Dense!   26
 11   Red Sky at Night   28
 12   A Growing Global Problem   30
 13   Acid Rain, Go Away   32
 14   Buoy, Oh, Buoy!   34
 15   Collapsing Under Pressure   35
 16   It's Just a Phase   36
 17   Falling for Color   38
 18   Fancy Footwork   40
 19   Fill 'er Up!   42
 20   Sound Advice   44
 21   Mining for Cereal   45
 22   Seeing Is Believing   46

23 Hero's Fountain   48

24 Drawing with Fire   50

25 Thanks for the Memories   52

26 Great Expectations   54

27 The Amazing Water Shooter   55

28 Here's Mud in Your Eye   56

29 Solar Sand   57

30 Any Which Sway   58

31 It's Elementary, My Dear   59

32 And They're Off!   62

33 Freefall!   64

34 Shape Up!   66

35 Fighting the Air Force   68

36 Here's Looking at You   70

37 The Big Crash   72

38 Beyond Explanation   74

39 "X" Marks the Spot   76

40 'Round the Bend   78

41 Magnetic Levitation   80

42 Half-Life Dating   82

43 Wayward Compasses   84

44 Spare Me the Details   86

45 Cosmic Ray Detector   88

46 Helloooooo!   90

47 Bubbles, Bubbles Everywhere   91

48 A *Cube-Shaped* Globe?   92

49 Only Your Cabbage Knows   93

50 It's Not Black or White   94

Standard Measurements and the Metric System   95

Index   96

Getting Started

Entering a science fair is an exciting experience. Perhaps that's because it's all up to you. You get to decide which project you want to do; you gather the materials you need and conduct the experiments; and, finally, you enter your project in the competition. With all the work you put into it, there's no doubt you would like your project to win!

But where should you begin?

THE SCIENTIFIC METHOD

The best place to start is with a very important tool called the *scientific method*. The scientific method is a set of ideas (a procedure) that scientists use to investigate things they want to understand. By using the method, you can be sure you're carrying out your project correctly.

The scientific method has five steps:

1. **Determining a scientific problem to solve/making an observation.** In other words, what's the purpose of your project? What question are you trying to answer?

2. **Developing a hypothesis.** What exactly is a hypothesis? A hypothesis is a speculation, a guess about how or why something happens. Based on your hypothesis, you can predict what outcome you expect for a particular experiment you do.

3. **Testing your hypothesis.** You must carry out some experiments to test your hypothesis. Finding out that your hypothesis is wrong is just as good as finding out that it is right. You must also check the results of your experiments against known facts.

4. **Recording your observations.** What did your experiments tell you?

5. **Drawing a conclusion.** The conclusion presents your *interpretation* of the results of the experiments you performed. If your experimental results agree with your original prediction, then they support your hypothesis. If not, you need to tell why you think your prediction differs from your results. Was there a problem in the way you did your experiment? Or do you need a new hypothesis to explain what you see?

Avoiding the Trap of the "Pet" Hypothesis

Sometimes it's very easy to just look for information that supports your best guess. For instance, let's say your project is "In which habitat are butterflies most likely to live, gardens or trees?" Perhaps you really like gardens, so you watch the butterflies in your flower garden. Your hypothesis, as a result, may be "Butterflies are most likely to live in gardens." You take this one step further by recording *only* the data that shows that butterflies live in gardens; you ignore or record only a little of the information that shows that they live in trees. If this is what you do, you are guilty of falling into the trap of the "pet" hypothesis—that is, a hypothesis you like. In reality, you should be looking for the answer, *whatever it is*. Proving your hypothesis is wrong is just as valid as proving it is right.

WHAT MAKES A GREAT SCIENCE FAIR PROJECT?

Science fair judges know exactly what they are looking for. Many projects they see are just poster displays of pictures cut from magazines or words copied from a textbook or encyclopedia. But these kinds of projects have very little to do with actually doing science. Doing science isn't assembling facts and writing reports. Instead, it's a process for exploring the unknown. The whole purpose of a science fair, then, is to get you involved in the exploration!

In a great science fair project, the students, just like explorers, leave a diary of their journey for others to follow. They leave the details of their exploration, from the reason they went (the scientific answer they sought) to the ideas they used to guide them (their hypothesis) to the experiences they had (the experiments they performed) to the final outcome of their journey (their conclusion). Such a winning project takes much more thought and time than a poster display or written report.

To be really great, a project must come from a creative imagination and a serious, scientific look into the real world. It should clearly show the problem that has been faced and the solution that has been discovered. In other words, it has to show that the student has used the scientific method. The best projects also tell an exciting story about how, with curiosity and hard work, the unknown became the known.

6

FOLLOW THE RULES!

When you enter a science fair, you have to follow certain rules. Your teacher can tell you what the specific ones are for your school. Here are some common rules that most science fairs require:

- All projects must use the scientific method (see page 5).
- Only one project is allowed per student, and only one student may work on each project. (The exception is a project done by two to four students from the same school who enter the "group" category.)
- The student is responsible for selecting an appropriate category in which to enter the project.
- Maximum exhibit size is 4′ wide by 2½′ deep by 6½′ tall.
- All projects must clearly distinguish between the work of the student and the work of others who have helped.
- All projects must be set up and picked up only during designated hours.
- Students must take full responsibility for the safety of all parts of their exhibits.
- Exhibit backboards must be able to stand up on their own.
- Any exhibits using electricity must be designed to use 100 volts and are limited to 500 watts (ask your teacher for more details).
- All exhibits must follow all county, state, and federal laws and regulations regarding wiring, toxicity, fire hazards, and general safety.
- Students must be available to speak to the fair judges at the designated time.

Most science fairs also do not allow projects that use live *vertebrates* (animals with backbones, such as frogs, fish, birds, and mammals). Check your science fair rules before considering using them. If you do use an animal, make sure that it is not harmed in any way. Animals that you can use include *invertebrates* (those without backbones), such as insects, worms, and crabs. (Some schools allow only *pictures* of invertebrates, not the animals themselves.)

Here's a list of items you cannot have at the science fair:

- preserved animals or their body parts
- human body parts
- microbial cultures and fungi
- food syringes
- pipettes
- hypodermic needles

- drugs
- dangerous chemicals
- materials that are combustible (easily catch fire) or explosive
- radioactive materials
- batteries with open-top cells

Electronic Devices

You can use all sorts of electric and electronic devices in your project. However, devices that involve wiring, switches, and metal parts that use 110-volt power or higher voltage must be located out of reach of your viewers and must be properly "shielded." That means that all moving parts must be covered and all wire connections must be enclosed so that no one can get hurt.

Some electronic devices you might use are:

- hair dryer
- electric fan
- cord and socket for an electric light

- ultraviolet lamp
- transformer
- computer

WHAT'S NEXT?

Once you've chosen your idea, you must develop a project around it. To do that, you must have a clear vision of what your project will accomplish. In other words, you must:

1. **Figure out the purpose of your project.** What is the scientific problem or question you are trying to answer? When you have that figured out, the scientific method comes into play. What observations will you be making? How will you develop a hypothesis? Make sure that you're following the scientific method every step of the way.

2. **Develop a plan for carrying out your project.** How much time do you have to organize your project? How are you going to gather your information? (Your project will be a lot more interesting if you can include some background information about it.) What kind of experiments will you conduct, and how much time will they take? Decide on a plan of action for every stage of your project, and write your plan down on paper.

WINNING TOUCHES FOR YOUR PROJECT

The little things are what capture the judge's eye. Sometimes they make the difference between a winning project and one that doesn't quite make it. What are those little things? Follow these tips to make your project extra special.

1. **Make sure your project is noticed.** Make sure you can see your project title from a distance. Also, make the lettering for your various headings large enough to be read from a distance. Perhaps you or a parent can make your titles on a computer—then you can make them any size you want. If you don't have access to a computer, you can buy press-on lettering from an art supply store.

2. **Display terrific artwork.** Choose your pictures carefully. Make sure the artwork is first class. Get an artistic friend or your parents to draw the pictures if they are too difficult for you to draw. Use tables, charts, graphs, and photos whenever possible.

3. **Make colored backgrounds for your pages.** Use colored paper as a background frame for each page of your report. Make sure you have no misspellings or mistakes anywhere. Glue everything to the backgrounds so that the glue doesn't show. Neatness counts!

NOTE

The numbered beaker in the upper right-hand corner of each project indicates the level of difficulty; 1 being the easiest, 3 being the hardest.

4. **Design a creative, colorful display.** Remember to keep your booth to a size that will allow everyone to comfortably read the papers hanging from it. Paint your display booth a bright color that contrasts nicely with your colored paper backgrounds.

5. **Keep your display clean.** Cover your entire display with clear plastic to keep the fingerprints off until the judges come through.

6. **Use working models.** If possible, use working models rather than stationary models. Then the judges and your viewers can see your project in action.

7. **Dress neatly.** Wear clean, neat clothing, so the judge's perception of you is that of a future scientist. You don't have to wear dressy clothes, but wear something nice.

8. **Stay with your project.** Stay with your project until after the judging is over. The judge(s) will want to ask you questions and so will your viewers.

Now that you've learned a little bit about the scientific method, the rules of science fairs, and making your project look its best, you need to choose the project that interests you most! The following are 50 very nifty science fair projects. Look on page 30 to test the greenhouse effect, page 52 to study peoples' memory-holding capabilities, page 16 to find out the importance of water in food, page 46 to perform your own consumer science testing, and 46 others.

MEASURING METRIC SHORTCUTS!

If you or your school use the metric system, on page 95 you'll find a handy metric conversion guide that converts the standard measurements found in this book into metric.

To change other measurements you come across during your research, use the simple table below.

If you know . . .	You can find . . .	If you multiply by . . .
inches	millimeters	25.4
feet	centimeters	30.0
pounds	kilograms	.45
ounces	milliliters	30.0
pints	liters	.47
quarts	liters	.95

The Heat Is On

PARENTAL SUPERVISION REQUIRED
If you put a paper cup above fire, it will quickly burst into flames. In this project, amaze your friends with paper that doesn't burn.

Materials

- paper cup (unwaxed)
- water
- metal rod and clamp (found in most chemistry labs)
- Bunsen burner

Procedure

1. Fill the cup nearly to the top with water. Hang it from the metal rod and secure it in place with a clamp.

2. Place it above the Bunsen burner. It should be at a height so that, when lit, the flame is about an inch below the cup (A).

3. With an adult's help, light the Bunsen burner, being careful not to let the flame directly touch the cup.

4. Watch it closely. The water will get hotter and even boil, but—amazingly—the paper cup will not burn.

Ⓐ

Analysis

When you boiled the water, you brought its temperature to 100°C, which is the *boiling point*. In order to have made the paper burn, however, it would need to reach the *kindling point* (the temperature at which paper catches fire). This is much higher than 100°C. Because of this, the heat is used to keep the water boiling because, to put it simply, this requires less effort for the flame than burning paper. Only once all the water evaporates will the paper cup finally catch fire.

Get the Point?

PARENTAL SUPERVISION REQUIRED

If someone laid down on a sharp nail, it would really hurt. But did you know it's possible for a person to lie on a *bed* of nails, and not get hurt? With this project, you can show how this is possible, using a balloon instead of a body.

Materials

- cardboard box about 1' square
- serrated knife
- hammer
- five nails, all the same length
- two wood boards, each 1' square, approximately ½"–1" thick
- balloons (about 8" in diameter when inflated)
- colored markers (optional)
- plastic bowl
- 5 lb. bag of sugar
- black marker

Procedure

1. Remove the top of your box. Use a serrated knife to cut off the front side and keep this piece for later.

2. Hammer a nail completely through the center of one of the boards so the head is pounded flat and the pointy end pokes out the other side. On the other board, pound in four nails so they form a 2" square in the center (A). They should all be the same height.

3. Place the board with one nail inside the box. Now blow up two balloons and set one lightly on top of the nail. (If you want to "drive home the point" that the balloons represent people, go ahead and draw funny faces on them.)

4. Remember that piece of cardboard you saved? Set it on top of the balloon. It should rest flat like a shelf, just barely grazing the sides of the box. Now set a plastic bowl on top of the cardboard (B).

5. Pour sugar very slowly into the bowl. You'll notice that the weight of the sugar causes the shelf to lower, pushing the balloon onto the nail. When the balloon pops,

use a marker to record on the inside of the box the point that the shelf was at when the balloon burst.

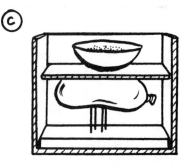

6. Now empty the bowl and replace the board with the one with four nails. Place another balloon on the nails, cover with the piece of cardboard, and set the plastic bowl on top (C).

7. Slowly pour sugar into the bowl again and use a different colored marker to record on the box when the balloon pops. Did the balloon handle more weight this time?

Analysis

With your balloon experiment, the nails were all equally strong. There was one critical difference, however, and that was *shared distribution of weight*. With just one nail, its tiny tip alone had to handle the pressure of the sugar's weight. No wonder your balloon probably popped quickly! When you added more nails, that meant that not one but four were sharing the burden of the weight. Each nail had to do only one-fourth of the job that the single nail did. The result is that more weight could be applied before the balloon popped.

Shared distribution of weight also explains how a person could lie on a bed of nails without looking like a human pincushion, provided there were enough nails to each hold a tiny portion of the person's weight.

The "Eyes" Have It

PARENTAL SUPERVISION REQUIRED

If you slow down a cartoon, you'd see that it is simply a series of illustrations. So what makes the drawings seem so real? Filmmakers rely on our *persistence of vision* to add life to still images. In this project, you can put on your director's hat and perform a bit of moviemaking magic yourself.

Materials

- white posterboard
- scissors
- ruler or T-square
- pencil
- large paper clip
- utility knife
- glue
- white paper
- colored pencils
- lazy Susan

Procedure

1. Cut a strip of posterboard measuring 2′ by 5″. Lay the posterboard in front of you horizontally and make a small pencil mark near the top of your strip every 2″ from the left side to the right. Use a ruler or T-square to make sure your measurements are precise.

2. Set the paper clip vertically over your first dot so its tip is ¼″ from the top of the strip. Trace around it and repeat this step on each mark. With a parent's help, use a utility knife to carefully cut out the paper clip shapes (A).

3. Glue the ends of the strip together to form a cylinder and place it on the lazy Susan.

Ⓐ

4. Now cut a strip of paper 3″ by 2′. Use a pencil and ruler to divide it into eight squares, each 3″ wide.

5. Ready for a little artistry? You're going to draw a simple action, breaking it up into eight simple steps—one for each of the boxes on your paper strip. For example, to draw a kite sailing up in the air, start with a kite on the lower left corner in the first box, then show it slightly higher in each box until it's high in the frame on the last box. You could draw a ball bouncing, a person dancing, or an ice-cream cone melting—whatever you want.

6. Slip the cartoon strip inside the cylinder with the images facing toward the center. The strip will rest just under the holes you cut in the cylinder (B).

7. Ask someone to crouch so he or she can see the drawings when peeking through the holes in the cylinder. Then spin the lazy Susan. Your cartoon will appear to come to life!

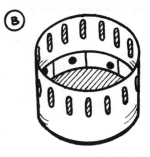

Analysis

"Seeing" is not simply taking a picture of what's really around you. What you see is a combination of a few things: light reflected from (or given off by) the objects around you, images your brain expects to see, and traces of things your eyes have just been exposed to seeing. This project takes a look at the last of those possibilities: Why do you continue to envision something you just saw? Scientists know that when light entering your eye reaches the light-sensitive lining (the *retina*) at the back of the eye, small cells in the retina respond. They pass messages along to the optic nerve of the brain, where you "see" a picture. While this process all takes place in an instant, your brain continues to see a picture one-tenth of a second after it has disappeared. This is referred to as *persistence of vision* or, if what you're staring at is particularly bright, as an *afterimage*. For example, stare at a lightbulb, then turn away to look at a wall: Does the image of the lightbulb seem to move to the wall? The same effect applies to your homemade cartoon. Since the cylinder is spinning quickly, your brain overlaps images, causing them to appear to move.

Water Weight Gain

PARENTAL SUPERVISION RECOMMENDED

All foods contain water. Can you tell how much water is in something just by looking at it? Let's find out!

Materials

- various foods, such as cabbage, watermelon, carrot, potato, bread, celery, and cereal
- knife
- balance scale
- paper clips
- pencil
- paper
- cookie sheet
- oven mitts

Procedure

1. Select several items of food. Where appropriate, cut the food so you have a chunk no larger than a tennis ball. For example, you might want to use a whole slice of bread but cut out just a hunk of cabbage.

2. Weigh each item on your balance scale. Use paper clips to provide the "balance," and on a piece of paper record the number of paper clips each item weighed before your experiment (A).

3. Set the pieces of food on a cookie sheet, spreading them out as much as possible.

4. How much water do you think is in each piece of food? Write down your guesses using percentages (10 percent water? 20 percent? and so on).

5. With an adult's help, place the cookie sheet in an oven at low heat (about 200 degrees). Let it bake for 2 hours, then remove the pan. *Remember to use oven mitts!* Re-weigh each food item. You'll notice that some foods lost weight. That's because some of the water in them evaporated in the warm oven. Record any changes, then place the cookie sheet back into the oven.

6. Repeat step 5, checking the food every half hour, until you no longer find any changes in the food's weight, which means that all or most of the water has been removed.

7. How much water did the food contain? To find out, weigh each item one more time. Then divide your original findings by the current weight. If the bread weighed eight paper clips to start and now weighs six, it is six over eight, or three-fourths, its original weight. That means one-fourth of its weight—25 percent—was water.

8. Compare these results to the guesses you made in step 4. Did the foods that looked "wettest" actually contain the most water? Did foods have more water than you'd thought? Did they have less?

Analysis

Water is essential for life, and all living things contain a lot of water. Although you might live for two or three weeks without food, without water you could only survive a few days. All you need to do to see how difficult it is to actually *see* water is to look in a mirror. Amazingly, the human body is made up of 70 percent water! This includes water from solid foods as well as from beverages.

Some animals get all of their water from solid food, but humans normally need to drink about 1.06 quarts or 1 liter per day.

Are Your Suds "Duds"?

Can you get better suds without changing shampoos or using half a bottle each time you wash your hair? If your lather tends to lack luster, maybe the problem is that your home has *hard water*. With this project, you can find out why it's not easy to make soapsuds or lather with hard water.

Materials

- liquid soap
- distilled water
- masking tape
- marker
- one screw-top jar for each type of water you're testing plus one extra jar

- tap water
- water from other sources, such as a river, neighboring cities, lake, rainwater
- eyedropper
- pencil
- paper

Procedure

1. Mix equal amounts of liquid soap and distilled water to make a soap solution. (Distilled water contains no salts, which are what make water "hard.") Label the jar "soap."

2. Pour distilled water into a jar so it's half full. With a piece of masking tape and a marker, label the distilled water "distilled." This is your control jar—that is, the standard to test your results against.

3. Fill each of the other jars with an equal amount of each type of water you are testing. Label each jar according to the water you put in (A).

4. Using the dropper, put one drop of soap solution into the distilled water, screw on the lid, and shake it. Add one drop at a time until you make a foam (B).

5. Make a note of how many drops of soap solution you needed to make the distilled water foam. Then see how many drops you need to make each of the other types of water foam. The more drops of soap solution the water type needs, the harder it is.

Analysis

Because underground water is in contact with soil and rock particles, it usually contains calcium, magnesium, or iron. Water that contains a lot of these minerals doesn't form suds easily and is called hard water—perhaps because it is so hard to get a lather going! Many cities put additives in their water to soften it, or homeowners can purchase special water-softening equipment. You don't want your water to be too soft, however, because softer water forms so many suds it makes rinsing difficult. More importantly, some minerals in hard water can make it healthier to drink than soft water.

Earn a Good Conduct "Metal"

PARENTAL SUPERVISION RECOMMENDED

Sometimes when you pick up a spoon from a saucepan, it's so hot it burns your fingers. That's because heat travels from the pan through the spoon, a process called *conduction*. Do all materials conduct heat at the same rate?

Materials

- water
- teakettle
- butter
- items made of a variety of substances to test, such as a wooden spoon, metal spoon, plastic spoon, drinking straw, glass stirrer, twisted copper wire, and brass letter opener
- plastic buttons
- beaker
- aluminum foil

Procedure

1. With an adult's help, heat up some water in a teakettle on the stove.

2. Stick a small dab of butter as close as possible to the top of each test item. Each dab should be the same size and an equal distance from the bottom of each item (A). (So if you lined up all the items, the butter would form a horizontal line.)

3. Place a button on top of each dab of butter.

4. Pour about 3" of hot water in the beaker, and cover the beaker with aluminum foil to keep the steam from melting the butter. Then poke each item through the foil and stand it upright in the beaker.

5. Heat from the water will be conducted upward through the items. When heat reaches the butter, it will melt and the buttons will fall off. But since the materials conduct heat at varying rates, you can have a race to see which button falls off first. The item whose button falls off first is the most effective heat conductor (B).

Analysis

Each of your test items is made up of particles (atoms, ions, or molecules) that are in constant motion. The energy found in these constantly moving particles is called *kinetic energy*. However, not all particles move at the same speed. The total amount of internal kinetic energy an object has is referred to as its *thermal energy*. Heat is the amount of thermal energy that an object is able to transfer to another object. When you heat the bottom ends of the test items, thermal energy from the water causes the particles to speed up or increase in kinetic energy and travel through the items more quickly. Heated particles move more easily through some kinds of materials—such as aluminum and copper—than through others, so these materials, the better conductors, heat up and melt the butter first.

Hot, Hot, Hot!

You may have noticed how on a hot sunny day, you often feel cooler when you wear light-colored clothes, especially white. Is it just your imagination, or do light colors really keep the sun's rays away from you?

Materials

- Popsicle™ sticks
- glue
- scissors
- black and white paints and paintbrush
- masking tape
- four identical, small outdoor thermometers

Procedure

1. You'll use the Popsicle sticks to build three identical houses. To do this, first make a frame for each house by laying four Popsicle sticks in the shape of a square, holding them together with lots of glue. Make another square the same way, then connect the two squares by gluing a stick vertically from each corner of one square to each corner of the second square (A).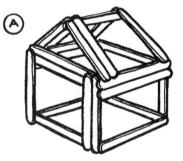

2. You can leave your roof flat, or if you're feeling creative, make a sloped roof like you see (you'll have to use strong scissors to cut some of the Popsicle sticks to fill in the gaps on the roof).

3. Once your frame is finished, glue sticks "log cabin style" along every side and on the roof. Make sure the houses are completely identical and sealed on all sides (except the bottom).

4. Paint the outside of one house black and paint another white. Leave the third house plain.

5. Glue or tape a thermometer in the same location inside each house.

6. On a sunny day, set the houses outside. Make sure they're all on the same sort of surface, such as grass, and they're all facing the same direction. Lay one thermometer next to the houses (B).

7. After an hour, compare the temperatures inside each of the houses. Is any house hotter than the others? How do the temperatures in the houses compare with the outdoor temperature?

Analysis

When light photons from the sun's rays strike a surface, some are absorbed while others are reflected, which is called *reflectivity*. The absorbed rays make you hotter, while those reflected don't affect temperature. To understand how this happens, you actually need to think backward. A white shirt doesn't reflect rays because it's white—it's actually white *because* it reflects rays. Sound crazy?

Here's how it works: The color of an object is determined by the frequencies of the sun's light rays (radiation) that it reflects. If all visible rays of the spectrum are reflected, the object appears white. If they're absorbed, the object appears black. Other colors are determined by the number of rays—and which rays—are reflected or absorbed. Reflectivity is the reason your white house stayed the coolest, and also why you probably see very few homes painted black in your neighborhood!

Penny for Your Thoughts

How many drops of water fit on the head of a penny? There are special forces that exist at the surface of water (or any liquid) that result in *surface tension*, and it's because of this you may just be surprised at the results of this project.

Materials

- penny
- cup
- water
- eyedropper

Procedure

1. Lay your penny flat on a table and fill a cup with water.

2. Ask a volunteer to guess how many drops of water fit on the head of a penny. Based on the size of a penny versus that of a drop of water, most people will guess around five or ten.

3. Now have the volunteer start dripping water onto the penny with an eyedropper, and count the actual number (A). You'll be amazed at how many drops really fit on such a small surface!

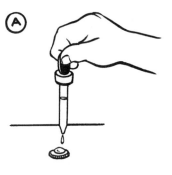

Analysis

Surface tension is the force that keeps drops of liquid such as water together. The molecules that make up the liquid are pulled toward each other, so molecules on the surface are pulled back into the liquid. This causes the liquid to change shape to keep its surface as small as possible and gives it the effect of an elastic skin, known as a *meniscus*. This is why drops are round.

When a drop of liquid touches a solid, the shape that the drop takes depends on what kind of surface the liquid is on. If the liquid molecules are not pulled toward the solid—such as when water comes into contact with a smooth surface such as a penny—the drop will remain as a drop. If, however, the liquid molecules are strongly attracted toward the solid—such as with paper or cloth—the drop will spread out and be absorbed (B). (See also Shape Up!, on page 66.)

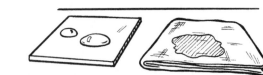

Water drop keeps its shape on piece of glass.

Water drop is absorbed into cloth.

Oh, Grow Up!

Plants will often change their usual growth patterns in low light conditions, even creating a series of twists and turns to seek out a life-giving source of sunlight. But what impact does gravity have on how plants grow? If a seed is planted sideways, will it know which way is up? This project will shed new light on plant growth.

Materials

- corn seeds
- water
- clean, empty jar with lid
- nonabsorbent cotton
- white paper towels
- clay

Procedure

1. Soak corn seeds in water for two days. Then arrange them inside the bottom of a jar so that they are in a circle, as though it were a clock with a seed at noon, and three, six, and nine o'clock. Make sure the wide side of each seed rests against the edge of the jar and its tip points inward.

2. Place a thin layer of nonabsorbent cotton over the seeds. Fold a white paper towel so it fits inside and lay it on top of the cotton to hold everything in place (you can use more than one towel if you need to make sure the seeds are held snugly). Wet the paper towel thoroughly, then seal the jar with its lid.

3. Place the jar on its side in dim lighting. It's important that the lighting be equally dim on all sides so that light doesn't affect the results of your experiment. Rotate the jar so that one kernel is pointing straight up. Use bits of clay on each side of the jar to prevent it from rolling (A).

4. Observe your experiment every day for a week. Make notes and draw what you see. Do different parts of the plant react to gravity differently? Do the roots and shoots always grow from the same part of the kernel? How long does a plant take to respond to gravity? Is it the same for all parts?

Analysis

When a plant grows in a certain direction in response to a stimulus, it's called *tropism*. In this project, you observed *geotropism*—when a plant grows a particular way in response to gravity. A plant also may grow in response to light (*phototropism*) or water (*hydrotropism*). Try to imagine what would happen if geotropism didn't exist. Farmers would have to hand-plant each seed to ensure it was facing up. Otherwise, plants wouldn't grow up to seek the sun or send roots down for water.

Don't Be So Dense!

Have you ever heard the saying "oil and water don't mix"? That's because these liquids have different *densities*—that is, one is heavier than the other. What happens when you try to mix liquids of different densities? How will they respond if you try to add solid objects? This project will make the concept of density appear crystal clear.

Materials

- honey
- two beakers, each at least 8" tall
- drinking cup
- food coloring
- water
- vegetable oil
- two each of small objects of varying weight, such as tiny feathers, paper clips, marbles, nuts, pennies, birthday candles, and dry noodles

Procedure

1. Pour honey into a beaker (beaker A) so it's 2" high on the sides. Be careful not to let any drip down the inside of the glass.

2. In a drinking cup, add a few drops of food coloring to the water. Slowly pour the colored water on top of the honey so it, too, is 2" high (A).

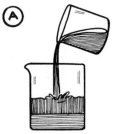

3. Add the same amount of oil to the beaker (2" high).

4. Pour identical amounts of the same liquids into the second beaker (beaker B). This time, however, add them in reverse order (first oil, then water, then honey).

5. One by one, drop each of the solid objects you selected into your two liquid cocktails (B). Make sure that you drop in the same objects at about the same time.

 Can you guess what will happen to the objects in beaker B?

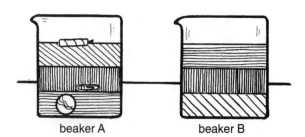

beaker A   beaker B

6. What happens to the objects in beaker A after ten minutes? After an hour? After 2 hours? Write down how far each object dropped or whether it was "stopped" by any liquid.

7. At the same time intervals listed in step 6, observe the changes in beaker B. What changes are occurring among the liquids? How is that affecting the solid objects?

Analysis

The *density* of a material decides whether it will sink or float. Materials denser or heavier than a liquid will sink in it, and even a liquid that is denser than another will "sink." Materials that are less dense than the liquid they are in will float. The density of an item is determined by a combination of the weight of its atoms, and also how closely these atoms are packed together. For example, even though an atom of lead is nearly eight times heavier than that of aluminum, lead is only four times denser because the atoms aren't as crowded together.

Many of the items in your first beaker got stuck at a layer of liquid simply because they're less dense. Did the liquids and solids in your second beaker try to arrange themselves in order of density? Were they completely successful? If not, it's probably due less to density and more to Newton's First Law, which states that an object at rest will remain at rest unless acted upon by an outside force.

Red Sky at Night

Have you ever wondered why the sky turns orange and red when the sun sets? In this project, you can demonstrate why—in full color!

Materials

- water
- long, clear plastic dish
- flashlight
- stack of books
- spoon
- milk

Procedure

1. Pour water nearly to the top of a long, clear plastic dish.

2. Prop a flashlight on a stack of books so it's shining through your dish lengthwise (A).

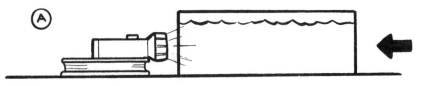

3. Stand at the opposite side of the dish and peer toward the light. You'll see that it looks white.

4. Add a few tablespoons of milk to the water, then stir the milky water with a spoon.

5. Now stand at the opposite side of the dish again and look through the milky water. You'll see that it appears to be lit with an orange-red color!

6. Dump the water and repeat the experiment. This time, stand to the side of the flashlight rather than at the far end of the dish when you look into the water (B). Before milk is added, the water will appear clear. Afterward it will appear blue.

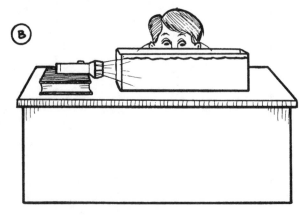

Analysis

Sunlight is not just one color but a mixture of many colors. In air, all the colors move at the same speed and stay together. However, back in the 1600s, Isaac Newton discovered how light can be split by a triangular piece of glass called a *prism*, which slows down the speed of light. Colors lose speed at different rates, so the rays actually bend at different rates. Red rays are the "fastest" and violet rays are the "slowest." When light from the sun passes through the air, tiny particles of dust get in the way of the colors. Reds and blues split in different directions. That's why when you looked directly at the flashlight—as though you were looking at a sunset—the speedy red rays made their way straight through the specks of milk. When you looked at the side of the dish, you saw the blue color that was scattered by the specks.

A Growing Global Problem

Without using a heater, a greenhouse can grow plants that wouldn't be able to survive outside. How does a greenhouse work? What's the big deal about the "greenhouse effect"? Find out with this simple project, which must be done on a sunny day.

Materials

- two identical plastic cups
- water
- glass bowl
- two thermometers
- paper
- pencil

Procedure

1. Fill each cup nearly to the top with water.

2. Put a glass bowl upside down over one of the cups and leave the cups in the sun for about an hour (A).

3. Remove the bowl and take the temperature in each cup using your thermometers. Write the results down on a piece of paper so you'll remember them later.

4. Was there a difference in temperature between the two cups? If so, divide the temperature of the warmer into the cooler, and multiply by 100. Then subtract the number from 100 to get the percentage difference. For instance, if the temperature in the first cup was 75 degrees, and the second cup was 60 degrees, the difference is that the first cup is 20 percent warmer.

Formula:

100 − (cooler temperature/warmer temperature × 100) = % difference in temperatures

Analysis

In your experiment, the heat rays of the sun passed easily through the glass bowl and heated the water in the cup. In a greenhouse, the sun causes the water exhaled by the plants to condense into water vapor. The water vapor gives off heat, but at a lower temperature than that of the sun. This heat cannot pass back outside the glass, and the space inside the greenhouse heats up.

While greenhouses can help plants thrive, something called the "greenhouse effect" is a growing global problem. Think of the cup under the glass bowl as the earth, and the bowl as the atmosphere, which is protected by an invisible layer called the *ozone*. The ozone keeps out the most harmful rays of the sun, the ultraviolet (UV) rays. The earth then reflects much of the heat energy from the sun back into the atmosphere (B). A problem arises when much of this reflected radiation cannot escape because gases, such as carbon dioxide released by burning fuel, absorb the radiation and trap it inside the atmosphere, increasing the temperature of the earth. This is the greenhouse effect. Many scientists think that it may change the world's climate over the next hundred years or so. A global rise in temperature could eventually melt glaciers at the North Pole that are big enough to flood the entire planet! That's why many environmentalists are urging people to use less fuel by driving less, turning off lights, and conserving energy in other ways. Environmentalists also support the preservation of the rainforests, which help to keep the earth cool.

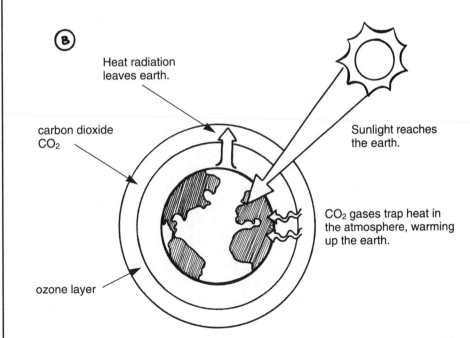

B

Heat radiation leaves earth.

carbon dioxide CO_2

Sunlight reaches the earth.

CO_2 gases trap heat in the atmosphere, warming up the earth.

ozone layer

Acid Rain, Go Away

You may have heard about an environmental problem called *acid rain*. Although all rain is acidic, some rain is so acidic that it can damage lakes and rivers, killing fish, plants, and other life. In this project, you'll witness firsthand its effects by creating acid rain for a pond filled with plankton.

Materials

- hanger
- nylon stocking (the single leg kind—if you use pantyhose, you'll have to cut off one leg first)
- stapler
- scissors
- glass or plastic vial, about 3" in diameter
- duct tape
- rope
- magnifying glass
- two glass containers
- vinegar or lemon juice
- pH paper (available from biological supply companies)
- pencil
- paper

Procedure

1. Make a ring 8" in diameter from a coat hanger. Stretch the thigh of the stocking around the ring and staple it in place.

2. Cut the toe from the stocking, then slip a vial just inside the stocking with the vial opening into the stocking. Hold it in place by wrapping duct tape all around the line where the stocking and vial meet.

3. Tie a rope to the wire ring to serve as a towline (A). You will need to cut a tiny hole on opposite sides of the nylon hose to pull the rope through.

4. To collect plankton, microscopic organisms, go to a stream, river, pond, tide pool, or lake. Drag the tow net in water, while walking along the shore or a dock.

5. Plankton is not visible to the naked eye. Use your magnifying glass to add equal amounts of plankton to each of your glass containers, then fill each nearly to the top with additional pond water (or from whatever type of water you pulled the plankton).

6. Set your containers side by side, leaving one as a "control" so you have something against which to compare the results of your tests.

7. Identify some of the organisms in your miniature pond. It may help to check out a book on microscopic sea life from the library. Draw pictures of what you find, or list them by type.

8. Each morning for a week add a tablespoon of lemon juice or vinegar to the second container (B). (These liquids are high in acid and will mimic the effects of acid rain.)

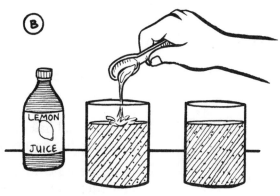

9. At night, test the precise pH balance of your mini-ponds. To do this, dip a pH strip into each container. (A strip of pH paper measures how many "parts Hydrogen" something has.) Then follow the directions on the package of pH strips to find out what the actual pH is.

10. Record your results each day. Also, write down your observations of any changes in your mini-ponds.

11. After one week, compare your acidic pond to the control container. Determine the effects, if any, of the increased acidity. Which organisms are most acid-resistant over time?

Analysis

Acid rain is the result of air pollution. Acid rain happens when gases, which are rising through the atmosphere, encounter the moisture in clouds. The moisture triggers a chemical reaction and forms sulfuric and nitric acids. When rain falls from these clouds, it's acid rain. Acid rain not only harms tiny organisms, but also can make people sick, since we drink the water and eat fish from these contaminated waters. As we burn more fossil fuels such as coal, oil, and gasoline, more dangerous gases pour into the atmosphere. By nature rainwater is somewhat acidic and ordinarily has what scientists refer to as a pH range of 5.6 to 5.0. A pH reading is a measure of acidity—anywhere from pH 1 to pH 6.9 is considered acidic. The lower the number, the higher the acidity. Now, in some parts of the country, rain can have a pH of 4.2, which is 25 times more acidic than regular rainwater. Think of tiny organisms that live in ponds and streams: A shift of even one or two pH units could kill them, even wiping out entire types of plants.

Buoy, Oh, Buoy!

What makes a buoy float? According to a famous Greek mathematician and inventor named Archimedes, if something is partly underwater—such as a boat, an iceberg, or a floating oil rig—it's buoyed up with force equal to the weight of the water it has taken the place of. This project will help you understand the forces behind floating.

Materials

- nail
- hammer
- wooden block
- hanging scale
- hook
- drip tray
- bucket
- pencil
- paper
- string

Procedure

1. Drive a small nail into the block of wood (you'll need this so that the wood can be weighed later).

2. Hang the scale from a secure hook. Place a tray underneath to catch water that will spill.

3. Fill the bucket to the brim with water and hang it from the scale. Write down the weight of the bucket and water.

4. Take the block of wood and put it in the bucket of water (A). Some water will spill out of the container. Does the reading of the scale change?

5. Remove the wooden block from the water, being careful not to spill any water from the container. What is the weight of the container now?

6. Subtract the weight you found in step 4 from the original weight of the bucket and water measured in step 3. This is the weight of the block of wood.

7. Double check your findings by taking the container off the scale. Tie a piece of string in a loop to the nail in the wooden block and hang it from the scale to determine how much the block weighs (B). Was the measurement you reached in step 6 correct?

Analysis

Archimedes' scientific principle is that the buoyant force on something immersed in fluid is equal to the weight of the fluid displaced by that object. In other words, the weight of the wood block should be equal to the weight of the water that spilled out over the sides of the bucket. (For more on weight distribution, see Get the Point? page 12.)

Collapsing Under Pressure

PARENTAL SUPERVISION RECOMMENDED

You may have heard that something is "lighter than air." Yet the weight of air—known as *air pressure*—is a powerful force. In this project, you can test how mighty air pressure can be on a small scale.

Materials

- thin metal can with airtight cap
- ½ cup water
- hot plate or gas burner
- oven mitt
- cloth

Procedure

1. Thoroughly wash and dry the tin can. Check with an adult to make sure the can that you have selected is safe to use—capped tin cans are often used to hold flammable materials, so you must make sure yours is thoroughly cleaned.

2. Pour ½ cup of water into the can. Heat the tin can over the hot plate or gas burner (make sure that the cap is off). Let the water boil for about 2 minutes, until you see steam coming out of the can (A).

3. With an oven mitt and the help of an adult, carefully remove the can from the heat.

4. Immediately close the cap very tightly using a cloth to protect your fingers.

5. Let the can stand upright on a table. Then watch it slowly crumple as it cools (B).

Analysis

Think of air as a constantly pushing force. In this project, you created an unusual situation where air pressure came in contact with *lack* of air pressure, and you demonstrated how air is strong enough to crush metal.

Here's how it happened. Originally, air pressure inside the can was the same as air pressure outside. When you boiled water, steam pushed air out of the can, which was trapped outside when you sealed the can. The pressure of the steam was still equal to outside air pressure. As it cooled, however, the steam condensed back into water. Since water used up less space than the steam, it caused the air pressure inside the can to drop. Pressure outside was then so much greater than the pressure inside, it crushed the can.

It's Just a Phase

Materials

PARENTAL SUPERVISION REQUIRED
Look up in the sky at night and the moon may appear to be anything between a crescent-shaped sliver and a bright sphere. These are known as the "phases" of the moon. What causes them?

- glue
- black paper
- shoe box with lid
- masking tape
- Styrofoam™ ball (about 2″ in diameter)

- black thread
- small flashlight
- pencil
- utility knife

Procedure

1. Glue black paper inside the shoe box and lid so that they are completely covered.

2. Use a small piece of masking tape to attach some black thread to hang the Styrofoam ball. Tape the other end of the black thread to the center of the lid, hanging the Styrofoam ball about 1″ down (A).

3. Lay the head of your flashlight against one of the two short ends of the box, and use a pencil to trace a circle around it. With an adult's help, use a utility knife to cut out the hole. Set the flashlight so that it fits into the hole and is directed inside the box. On the outside of the box, seal any space between the flashlight and the box with masking tape so that no outside light can leak into the box.

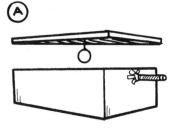

4. Cut five eyeholes in the sides of the box (B). Two should be on each of the long sides, and one at the foot of the box, located just beneath the flashlight.

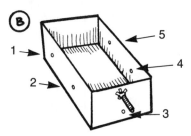

5. Set the lid on the box and seal all edges with masking tape.

6. With the flashlight off, have someone look in the box. He or she will just see blackness.

7. Now turn the beam on. Have the participant look first through eye-hole 1, then through eyeholes 2, 3, 4, and 5 (B). He or she will see the phases of the moon in the proper order.

Analysis

Half of the moon is always in sunlight (C). The phases of the moon de-pend on how much of the lit half we can see from earth. Because the moon turns on its axis at the same rate it revolves around the earth, the same side of the moon always faces earth. At the new moon, when the earth, moon, and sun are roughly in alignment, we can't see any of the lit half. About a week later, at the first quarter, we can see half of the part of the moon that is in sunlight. At full moon, we can see all of it. By the last quarter, we can again see only half of the lit part of the moon. In your project, the positions of the flashlight and the ball allow you to see the different phases because the alignment is similar to that of the sun, moon, and earth.

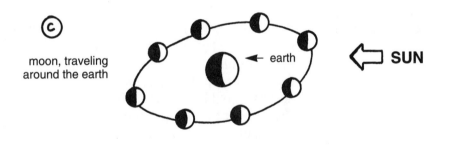

ⓒ

moon, traveling
around the earth

← earth   ⇦ SUN

Falling for Color

PARENTAL SUPERVISION RECOMMENDED
(*WARNING: Rubbing alcohol is extremely flammable, so keep it away from any flame or match.*)

As autumn arrives, leaves in many parts of the country take on brilliant tones of red, orange, and yellow. Why do leaves change color? How can a leaf be green one day, then suddenly switch to a whole new hue? In this project, you can demonstrate the answer.

Materials

- green leaves from trees, such as maple or elm, that change color in the fall
- filter paper or white blotter paper (available from science supply stores)
- coin
- rubbing alcohol
- several beakers, one for each type of leaf

Procedure

1. First, transfer a sample of the leaf tissue onto the bottom portion of your blotter or filter paper, which usually comes in small strips. (If yours doesn't, you'll have to cut it into strips about the size of Band-Aids™.) To make your leaf sample, place a leaf on the edge of the blotter paper. Then rub the edge of a coin over the leaf to crush it and transfer the green color onto the paper (A). Repeat this several times using a fresh part of the leaf each time so that you have a thick stain of color on the paper.

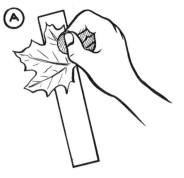

2. Prepare one strip like this for each type of leaf you collected. Let the strips dry.

3. Pour about an inch of rubbing alcohol into each beaker. Place one blotting paper leaf sample in each of the beakers (B). Watch what happens over the next few minutes. What do you see?

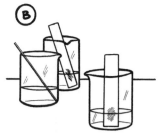

Analysis

Green plants make their food by a process called *photosynthesis*. There are several stages in photosynthesis, but the end result is that plants combine water from the soil with carbon dioxide from the air to make glucose (sugar). The process takes place only in the presence of sunlight and *chlorophyll*, the green color matter in plants. In the fall, just like some animals prepare to hibernate, some trees store nutrients in their trunks for the coming winter. To help the tree do this, the leaves stop pulling nutrients from the tree for photosynthesis, which over time causes the glucose and green-colored chlorophyll to get used up in the leaves. In this project, the green color is drained away through the chemical reaction between the alcohol and the filter paper. The colors left behind—such as red, yellow, and orange—seem to magically appear, similar to what you saw in your experiment. Leaves without nutrients can't hang on for long. That's why autumn leaves quickly wither and drop to the ground.

Fancy Footwork

Is it possible to do two simple tasks at the same time? In this project, you'll put your friends' coordination to the test.

Materials

- paper and pen
- tall table or surface to write on
- at least ten volunteers

Procedure

1. Set up a piece of paper and a pen on a table that a person could write on comfortably while standing.

2. Select a mix of volunteers who are right-handed and left-handed. (For your left-handed volunteers, reverse all the directions below.)

3. Have your volunteer hold a pen in his or her right hand. Tell him or her to rotate his or her right foot in a clockwise direction. While the foot is rotating, have him or her write a large number 6 on the paper (A). Watch the rotating foot to see which direction it moves in as your volunteer is writing, clockwise or counterclockwise. Write down your results (but don't tell the others what you've noticed . . . they may not even be aware).

4. Have your volunteer rotate his or her left foot clockwise, at the same time writing the number 6. Again, watch the foot. Does it try to change directions this time? Record the results.

5. Now ask your volunteer to rotate his or her right foot in the opposite direction, counterclockwise, while writing a 6. Note what happens.

6. Have your volunteer do the same while rotating his or her left foot counterclockwise, again writing down your results.

7. Repeat steps 3 through 6 with at least nine more volunteers so you have a large enough sample from which to draw a conclusion.

8. Graph your results. To do this, create a chart that lists each of the steps in the left-hand margin: right foot clockwise, left foot clockwise, right foot counterclockwise, left foot counterclockwise. Write

each volunteer's name in a row along the bottom of your paper, along with an "R" if they are right-handed or an "L" if they are left-handed. At the intersection where the volunteer's name and the task meet, make a check mark if he or she completed it successfully or write an X if his or her foot tried to change direction. Do you notice a pattern among your participants?

Analysis

Just like someone might be left- or right-handed, he or she might also be left- or right-brained! Coordination starts in a small part of the brain called the *cerebellum*, which receives its instructions from nerve impulses sent by the body. Here's where it gets tricky, however. The whole left side of the body is controlled by the right half of the brain, and the right side of the body is controlled by the left half of the brain. By writing a figure 6 with the right hand, the left half of the brain has instructed the right hand to make a counterclockwise motion (because typically a person writes the bottom loop of the 6 counterclockwise). The right foot could easily move in the same direction. The opposite movement requires a special effort because you're asking both parts of your brain to work at the same time.

Fill 'er Up!

Is it possible for a small person to have more lung power than someone who's much heavier or taller? In this project, you'll find out.

Materials

- graph paper
- pen
- lots of volunteers!
- balloons
- cloth tape measure
- red and blue markers

Procedure

1. Create a graph that you'll use to mark down your volunteers' results. On the left-hand side of a piece of paper, write "circumference in inches," and starting with 1" at the bottom, write numbers by ½" increments all the way until you reach 12" (A).

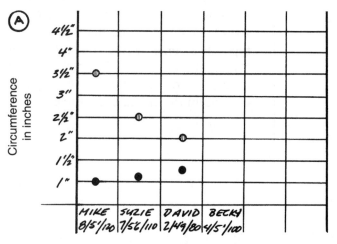

volunteer's name, fitness level, height, weight

2. Along the bottom of your graph, write your volunteers' names, their height, weight, and fitness level (see step 3). Try to get at least 20.

3. Find out your first volunteer's fitness level by asking how many minutes of strenuous activity he or she does each day. Rate their fitness level on a scale of 0 to 10, based on the following chart:

If a person exercises . . .	He/she rates . . .
1 to 2 hours, 5 to 7 days a week	10
1 hour, 4 to 5 days a week	8
30 min., 4 to 5 days a week	6
30 min., 2 to 3 days	4
30 min., 1 day a week	2
no exercise at all	0

Some of your volunteers won't exactly fit these scenarios, so try to find the fairest rating (some may be in between and will rate a 1, 3, 5, 7, or 9). Write each person's fitness rating on your graph just next to the person's name, along with his or her weight and height.

4. Determine the *reserve air capacity* of your volunteer by asking him or her to breathe out normally, then blow what air is left in his or her lungs into a balloon.

5. With the measuring tape, measure the circumference of the balloon (B). (Make sure you get the absolute center and widest part of the balloon to make your comparisons as accurate as possible.) Using a blue marker, place a dot on your graph where the name and the number on the left meet.

6. Now find the volunteer's vital capacity by asking him or her to take the deepest breath he or she can, then blow every bit of air into the balloon. Again measure the balloon's circumference and record the number on your graph using a red marker (A).

7. Repeat steps 3 through 6 with the rest of your volunteers. Compare your results. Did the heavier or taller people always have more reserve air capacity? What about vital capacity? Did physical fitness have a bearing on the results of either?

Analysis

When you get more exercise, your muscles need more oxygen, which they get when you breathe. Most people breathe in about a pint of air with each breath. A person may take in four quarts of air in a really deep breath—eight times the amount he or she would usually breathe. The more you "practice" breathing deeply, the more *efficient* your body gets at it. When you exercise, you bring a constant supply of oxygen to your lungs, which makes your breathing muscles stronger. To put it another way, it's not the size of your lungs that matters, it's how you use them that counts. You probably found in your project that people had similar *reserve air capacity*, which simply gives an idea of the size of their lungs. Their *vital lung capacity*—that is, the amount of air they took in when trying to fill their lungs as much as possible—most likely varied quite a bit. Vital lung capacity is what is increased by exercise. It can also be lowered by other factors, such as smoking, asthma, or breathing very polluted air.

Sound Advice

When we hear a sound, we're actually hearing a wave or vibration that has traveled through the air to reach our ears. What happens in places, such as outer space, where there is no air? Can astronauts hear in space?

Materials

- iron wire, approximately 2″ to 3″ long
- bell, small enough to fit in flask
- rubber stopper
- 1 cup of water
- flask with a flat base
- Bunsen burner
- cloth

Procedure

1. Attach one end of a small piece of wire to a bell. Firmly push the other end of the wire in a rubber stopper until it sticks in place. Adjust the length of the wire so that the bell hangs in the center of the flask when the rubber stopper is inserted (A).

2. Close the flask off with the stopper and bell. Shake the flask so you can hear the bell ring.

3. Take the stopper off and pour 1 cup of water into the flask. *Do not replace the stopper*.

4. Heat it over the Bunsen burner until the water boils, then let it boil for another minute before shutting off the flame (B).

5. Quickly insert the stopper so the bell is once again in the flask. (Hold the flask with a cloth so you don't burn your fingers.) Let the flask cool off for a minute or so, then gently shake it. Can you still hear the bell?

Analysis

A sound is simply a wave or vibration that is sent and received. If someone claps his or her hands, it sends a sound wave across the air. It then actually becomes a sound because you hear it in your ears. Sound waves can travel through gas, liquid, or solid, but they must travel through *something*. In this project, you removed the "path," in this case the air, that sound usually takes to reach a listener's ears. This was done when you boiled the water, replacing the air in the flask with steam, which was trapped with a stopper. As the flask cooled, the steam turned back into water, leaving nearly no air inside. The bell's ring couldn't be heard. Just as in outer space, there was nothing for the sound waves to travel through.

Mining for Cereal

Some foods naturally contain certain nutrients—for example, aspara-gus has a high content of the mineral iron. Can you detect if extra vita-mins or minerals are really in some foods? In this project, you'll go mining for iron in breakfast cereals.

Materials

- three bowls
- three breakfast cereals (check the labels: at least one should claim to be fortified with 100 percent of your recommended daily intake of iron)
- fork or blender
- three small magnets
- three wooden spoons

Procedure

1. Fill each of the three bowls about halfway with the different types of breakfast cereal.

2. Pour one bowl of cereal into a blender and crush it until it's as fine as sand. Pour the crushed cereal back into its bowl. If you don't have a blender, you can crush the cereal with a fork.

3. Repeat step 2 with the other bowls of cereal. Then place a magnet in each bowl of cereal (A). Swirl it around using a wooden spoon. (Use a different wooden spoon for each bowl to keep from mixing the cereals.) What happens to the cereal? Do all three cereals react the same?

Analysis

Iron is a trace mineral that is essential for the formation of *hemoglobin*, the red blood cells that give blood their color and enable them to carry oxygen. You get iron from food, in both green leafy vegetables and meats. You might be surprised to learn that the iron in these foods is the same iron found in the earth—the mineral that is used to make appliances or pots and pans! We don't often think of the two forms of iron as being the same, as you'd have a much easier time digesting an egg than the cast iron skillet it's cooked in! Food that is fortified with iron has had tiny quantities of this mineral added to it. In the case of the breakfast cereals, since it is possible to crush the flakes and break away the other ingredients, the bits of iron were pure enough to stick to the magnet.

Seeing Is Believing

Do products at the grocery store really deliver on the claims they make? Scientists are always seeking the "truth" in the universe, and that includes truth in advertising! In this experiment, you'll put paper towels to the test. Are some really stronger than others?

Materials

- scissors
- string
- two chairs
- three brands of paper towels, including one "bargain" brand
- clothespins
- unsharpened pencils
- notepaper
- pen

Procedure

1. Cut two pieces of string, each about 2′ long. Set two chairs with their backs facing each other, then tie the strings to the chairs so that the strings are parallel and about 10″ apart (A).

2. Thoroughly soak a sheet of one of your brands of paper towels. Lift it carefully and secure it with clothespins to the strings so that it stretches flat.

3. One at a time, lay pencils on the center of the paper towel. Keep adding pencils until the weight of the pencils causes the wet paper towel to break (B). Make a note of how many pencils it took to break the paper towel. Repeat step 2 and this step two more times, with the same brand of paper towel, and compute the average number of pencils that the towel can hold. To find the average, add together all three numbers and divide by three.

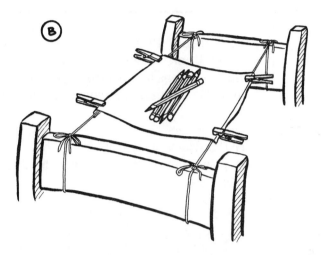

4. Now repeat steps 2 and 3 with the other two brands of paper towels.

5. Was there any difference in how much weight each could handle?

Analysis

Since the day Sir Isaac Newton "discovered" gravity by watching an apple fall from a tree (and probably even before!), scientists have been using the power of observation to challenge theories about the world. Depending on the brands of paper towels you chose, you may have discovered that the cheap brand was just as good as one more expensive—or you may have found out that the extra few pennies are worth spending. What you also learned, however, is how scientists constantly test what they have heard. When a scientist hears someone say "our paper towels are stronger," he or she says, "I'll have to see for myself." If scientists never questioned the *status quo*—that is, what everyone else believes is true—we'd probably all still think the world is flat.

Hero's Fountain

PARENTAL SUPERVISION RECOMMENDED
Do you think it's possible to create a fountain that flows all by itself, with no pumps to push the water? What is the secret behind what's called *Hero's fountain*, which can sometimes flow for 12 hours all on its own?

(NOTE: Many of these items will have to be made for you by a plastics or glass supply shop, or you may be able to borrow them from your school.)

Materials

- glass or plastic delivery tube (this is a bulb that has an open cavity on the top and a long stem on the bottom)
- two glass or plastic bulbs, 4" in diameter at the widest part, with 1" connecting ports on either end
- nozzle with a tiny exit hole (which should be a little larger than a pinhole), as well as a stem

- 10' of clear acrylic tubing, 1" in diameter
- silicone grease (available in science supply stores)
- wire cutters
- tie wire
- three ring stands with tube clamps
- water colored with food coloring
- table clamps

Procedure

1. Lay the delivery tube, bulbs, and nozzle on a table to figure out the lengths of tubing you will need to connect the parts. Then have the tubing cut to the appropriate lengths by a clerk at the supply store or by another adult.

2. Before sliding tubing over the small opening on a bulb or over the stems of the delivery tube and nozzle, smear a little silicone grease on the ends. This will make it easier to slide on the tubing (A).

3. Using wire cutters, cut some tie wire in pieces long enough to wrap around the connection ports and stems to hold the tubing in place.

Ⓐ

4. Once all the tubing is connected to the bulbs, the nozzle, and delivery tube, wrap a wire around each connection and pull it tight. This will help to prevent leaks.

5. Put together the end section and mount it in a ring stand. Fill the mounted end section with colored water (B). Later you'll fill the delivery tube with a different color of water.

6. Now assemble the rest of the fountain, connecting it to the end section as shown (C). *NOTE: The relative heights of the bulbs, nozzle, and delivery tube are critical, so closely follow the illustration.* Prop up the entire fountain with ring stands, tube clamps, and table clamps (D).

7. Pour colored water into the delivery tube until it is full and watch as your fountain begins to flow—all on its own!

Analysis

It may seem strange that water would travel up all on its own to form a fountain. This can be explained by air pressure. (For more on air pressure, see Collapsing Under Pressure, page 35.) When you poured water into the delivery tube, air was trapped between this section and the end section that you filled earlier. The trapped air forced water to flow from the nozzle. Gravity pushed the water down. As the water drained, the *volume* of air inside the tube increased—but the actual *amount* of air remained the same, as no air could get in from the outside. When the volume of gas increases in this way, its pressure decreases. As the pressure of air in the bulbs no longer pushes outward with the same force that air outside is pushing inward, the water shoots upward.

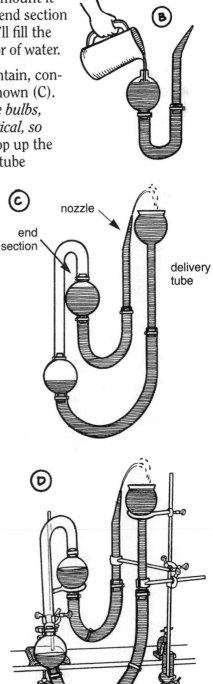

nozzle

end section

delivery tube

49

Drawing with Fire

PARENTAL SUPERVISION REQUIRED
This project experiments with chemistry. By starting an *oxidation reaction*, you can draw a picture with fire.

Materials

- glass of water
- small beaker
- glass stirring rod
- potassium or sodium nitrate (saltpeter)
- spoon
- paper
- pencil
- matches

Procedure

1. Pour about 1 cup of water into a beaker, then stir in about ½ teaspoon of saltpeter until it dissolves.

2. Keep stirring in spoonfuls of saltpeter until it no longer dissolves and starts to accumulate as a solid at the bottom of the beaker.

3. On a blank sheet of paper, draw an object on the paper using the glass rod dipped into your saltpeter solution. Make the lines fairly thick and use lots of the liquid (A).

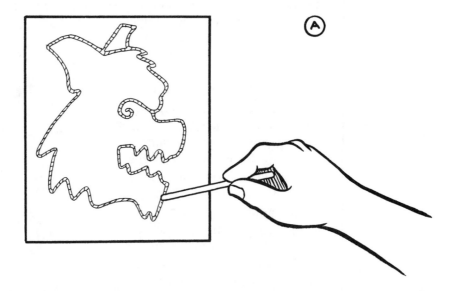

4. Leave the paper to dry. Before it is entirely dry, use a pencil to make a small mark where you started the drawing. The lines of liquid will disappear as your picture dries.

5. Show some observers your drawing, which will just look like a blank piece of paper.

6. Before you light the match, make sure an adult is nearby, as well as a glass of water. Strike a match, then blow it out. With the glowing tip, touch the marked spot on the paper (B). Immediately put the match into the glass of water to ensure that you have completely extinguished the flame. Then watch as your artwork appears to draw itself!

Analysis

Saltpeter is known as a strong *oxidant*. Oxidation is a process that takes place during chemical reactions when an atom or molecule gives up electrons, which causes energy. Oxygen atoms like those found in saltpeter take electrons from other atoms. In this experiment, once you drew a picture with the saltpeter solution, the water evaporated, leaving an invisible line of saltpeter (which, by the way, is one of the chief ingredients in gunpowder!). When you touched the glowing tip of the match, it started a chemical reaction, giving the first boost of energy needed to release the oxygen from the saltpeter. The active oxygen oxidized the paper, meaning it took the electrons and turned them into energy, which you saw as a burning glow. The energy was enough to keep releasing oxygen from the saltpeter, following the path of the drawing.

Thanks for the Memories

Short-term memory is the term for how we recall things we've had just a moment to consider. Have you ever forgotten someone's name just seconds after you were introduced? Then you know just how fleeting short-term memory can be. Is it possible to improve your short-term memory?

Materials

- variety of small, unrelated objects (ball, scissors, spoon, comb, CD, slipper, etc.)
- two trays
- volunteers
- several pencils
- paper
- stopwatch

Procedure

1. Arrange items on two trays so that each tray contains a different collection of objects (A).

2. Collect several volunteers, but don't let them see the trays!

3. Give each volunteer a pencil and two pieces of paper.

4. Set the stopwatch and show the volunteers one of the trays for 1 minute. Then remove the tray.

5. Tell them to make a list of what they saw. Collect their responses and make notes of how many items each person remembered.

6. Show them the second tray. Before you do so, however, give them tips for memorizing the objects. They can make "mental connections" between the objects. For example, volunteers may want to group items together by the letter they begin with or by their color. Another good memory booster is to make up a rhyme or story using the objects. Let your volunteers see the tray for a minute, then take it away.

7. Again have them make a list of what they saw. Collect their responses and make note of how many items each person remembered this time. Did they remember more?

Analysis

Memory is storing information to pull out later when you need it. There are several types of memory. The shortest type of memory is *sensory memory*, which lets you take in what you see, then instantly fades away. If you get a chance to think about something for a moment—as your volunteers did in this project—it becomes a part of *short-term memory*. Seeing objects only once will make them quickly disappear from short-term memory.

How are memories stored in the brain? Even with all the advances of science, no one really knows. Scientists think that memories involve chemical changes in groups of nerve cells and how these cells "connect" together or pass messages. The more these cells are used (meaning the more often you recall something), the better the cells pass messages and the more likely you are to remember something.

One thing is known about short-term memory: You can't improve it by practicing remembering things. Some people use tricks, such as rhymes, mental pictures, clues, and associations. You probably noticed in your experiment that your volunteers remembered more when they used these gimmicks. They didn't improve their short-term memory, however—just their ability to recall those few items.

Great Expectations

A famous scientist named Pavlov once showed how a *stimulus response action* could make dogs drool if they expected food—even if no food was served to them. Here, you'll test Pavlov's theory.

Materials

- at least ten volunteers
- roll of pennies for each volunteer
- a cup for each volunteer
- ruler

Procedure

1. Give each of your volunteers a cup and a roll of pennies. Have them unroll the pennies.

2. Stand so that you're behind  the group of volunteers. They should be able to hear you clearly. Tell your volunteers to drop a penny in the cup every time you say the word "drop" (A).

3. Strike a table with a ruler and at the same time say "drop." Do this exactly 20 times in a rhythmic manner, leaving 2 seconds between each time you say "drop."

4. Keep striking the table, but stop saying "drop." Continue until everyone has stopped dropping pennies. You only instructed them to drop 20 pennies. How many did most people drop?

Analysis

In this project, the word "drop" acts as a *response-eliciting stimulus* because it's directly telling people to do something. At the same time, however, you paired it with a *neutral stimulus*, which was the tapping of the ruler. A neutral stimulus is anything that gets our attention but doesn't directly give a command. Even though tapping the ruler had nothing to do with your instructions to drop pennies, your volunteers got used to hearing the two together. This is known as becoming "conditioned," meaning that they grew to expect the two things to happen together. Even when you no longer said "drop," many people probably continued to act as though you did. Conditioning is why the dogs in Pavlov's famous experiment started to expect food when they heard a bell ring—the scientist had sounded a bell every time he fed the dogs. After a while, people (as well as dogs) will notice that you're no longer giving them the response-eliciting stimulus (in this project, the word "drop") and they'll stop.

The Amazing Water Shooter

PARENTAL SUPERVISION RECOMMENDED

Someone who is wind sailing from a tall cliff will be carried much farther than someone who takes off from a tiny hill. Does the same theory apply to water? Will water shoot farther from the top of a cylinder than from the bottom?

Materials

- can opener
- large tin can (about 2 quarts)
- nail
- hammer
- pitchers
- water
- large washtub

Procedure

1. With an adult's help, use a can opener to cut off the bottom of the empty, cleaned can. Remove any label from the outside.

2. With an adult's help, hold a nail in your hand and hammer five holes in a straight line down one side of the can (A). Fill the pitcher with water and place the tin can in the washtub.

3. Ask a volunteer to cover the holes with his or her fingers, then pour the pitcher of water into the can (B). Tell your volunteer to remove his or her fingers. Out of which hole did water shoot the farthest?

Analysis

Water pressure depends on the "head" of water—the height at the top of the water supply above the level where it's used. Just as if heavy weights placed on water would push it out of a hole with great force, so does a heavy load of water. In fact, one of the reasons that divers are limited in how deep they can go is because they could be crushed by water pressure at too deep a level. In your experiment, since there was a much greater volume of water above the level of the bottom hole than there was above the top hole, the water stream at the bottom shot out the farthest.

Here's Mud in Your Eye

PARENTAL SUPERVISION REQUIRED

You wouldn't want to drink a big glass of mud. That's why many communities have water filters that remove dirt and other impurities before water ever comes out of people's faucets. In this project, you can make a water filter that—strange but true—uses sand and gravel to remove mud!

Materials

- knife
- 2-liter soda bottle with cap
- electric drill with ⅛″ bit
- plastic straw
- large clear vase
- cotton balls
- coarse gravel
- fine gravel
- large-grained sand
- fine-grained sand
- absorbent paper such as paper towels
- muddy water

Procedure

1. With an adult's help, use a knife to cut the bottom off the bottle, then drill a hole through the bottle cap just large enough to pass the straw through so that it fits tightly.

2. Place the bottle upside down so its neck rests in the vase. The straw should stick out through the bottom hole.

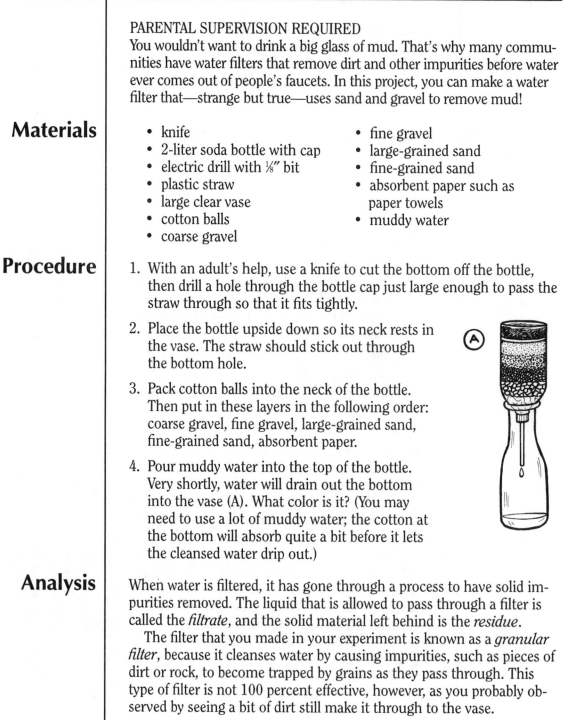

3. Pack cotton balls into the neck of the bottle. Then put in these layers in the following order: coarse gravel, fine gravel, large-grained sand, fine-grained sand, absorbent paper.

4. Pour muddy water into the top of the bottle. Very shortly, water will drain out the bottom into the vase (A). What color is it? (You may need to use a lot of muddy water; the cotton at the bottom will absorb quite a bit before it lets the cleansed water drip out.)

Analysis

When water is filtered, it has gone through a process to have solid impurities removed. The liquid that is allowed to pass through a filter is called the *filtrate*, and the solid material left behind is the *residue*.

The filter that you made in your experiment is known as a *granular filter*, because it cleanses water by causing impurities, such as pieces of dirt or rock, to become trapped by grains as they pass through. This type of filter is not 100 percent effective, however, as you probably observed by seeing a bit of dirt still make it through to the vase.

Solar Sand

The sun's rays on sand can make it too hot to walk on. Is it possible to heat up sand without any sun? In this project, you'll find that truth is sometimes stranger than *friction*.

Materials

- a medium-sized jar with lid
- dry sand
- thermometer (a cooking thermometer or one used for people will both work fine)

Procedure

1. Fill the jar about two-thirds full with sand. Push a thermometer into the sand.

2. After the thermometer has been in the sand for a few minutes, read the temperature of the sand.

3. Take the thermometer out of the jar, then seal it tightly. Shake the jar vigorously for about 5 minutes (A).

4. Open the jar again and push the thermometer into the sand, leaving it alone for a few minutes. Now what is the temperature of the sand?

Analysis

Solar power transforms heat into energy. In this case, you did just the opposite and turned energy into heat. *Mechanical energy* was required for you to shake the jar. This gave the sand particles *kinetic energy*, which is the energy of motion. This caused *friction* between the grains of sand. Enough friction can create heat, much in the same way that rubbing two sticks together can start a fire. That's why just a few minutes of shaking sand could actually raise its temperature. (Find out more about friction in And They're Off! on page 62.)

Any Which Sway

PARENTAL SUPERVISION RECOMMENDED

With this project, you can dramatically demonstrate the concept of *natural frequency*.

Materials

- drill
- piece of wood, 2″ × 4″ × 2′
- carpenter's glue
- six Plexiglas™ rods, all ¼″ in diameter, two 1′ long, two 2½′ long, and two 2′ long
- hammer
- six hard, brightly colored plastic balls, three 1½″ in diameter, and three 2½″ in diameter

Procedure

1. With a parent's help, drill six ¼″ holes in the wood.

2. Drop some glue into each hole, then tap a rod into each hole using the hammer. The rods should be in the order shown (A).

3. Next, drill a hole in each plastic ball. Put some glue into the holes and put one ball on the end of each rod, alternating the large and small balls.

4. Slide the wood back and forth on a table. The rods will start swinging—some violently, some hardly at all. *NOTE: Don't let them vibrate too wildly—the rods may break!*

Analysis

Think back to the last time you pushed someone on a swing. Do you remember that you pushed on the swing as it moved away from you, not as it came toward you? By matching your pushes to the natural movement of the swing, you added energy to the swinging and the swing went higher. Without realizing it, you were matching the swing's *natural frequency*.

What determines how fast or slow a natural frequency is? Well, with swings, the shorter the chains are, the faster the swing moves back and forth. But the bigger the person sitting on it is, the slower it swings. In your project, the balls on the rods are like kids on swings with chains of different lengths. When you shake the wooden board, little vibrations push on the rods. When these pushes match the natural frequency of a rod, it swings wildly. The long rods swing naturally slower than the short ones. When two are the same height, the one with the bigger ball on it swings more slowly.

It's Elementary, My Dear

When police find an unknown substance at the scene of a crime, they ask the police lab to study it. To discover what the substance is, the scientists and criminologists in the lab use *chromatography* to separate the different ingredients. In this project, you will get to be the criminologist as you solve the following crime:

The jewelry store has been robbed. The robber handed the jeweler a note demanding money. The note was written in black ink. The police have arrested four suspects, each of whom has a black pen. To identify the robber and solve the crime, you must analyze the ink in each pen and compare it with the ink on the note.

Materials

- four small white stickers
- four black ballpoint/marking pens of different brands
- three sheets of filter or chromatography paper 5" × 5"(available at chemistry supply stores)
- pencil
- scissors
- ruler
- cellophane tape
- wooden stick at least 9" long
- container at least 7" wide and 5" high
- water
- pencil holder
- four volunteers
- hair dryer
- paper towels
- 5½-oz. tin can with the bottom removed
- isopropyl alcohol
- large jar

Procedure

1. Number the stickers 1 through 4 and put a sticker on each pen.

2. On two sheets of the filter paper, draw a pencil line ¾" up from the bottom. Then cut the sheets into eight strips, each strip 1¼" × 5" (A).

3. Using the pencil, number each of the strips as follows: two labeled 1, two labeled 2, and so on through 4.

4. With the first ballpoint pen, draw an ink line over the pencil line on both strips labeled 1. With the second pen, do the same on the two strips labeled 2. Do the same for the remaining pens and strips.

(A) — diagram showing a 5" × 5" sheet marked into four columns numbered 1, 2, 3, 4, with the overall dimensions 5" wide and 5" high, and a pencil line ¾" up from the bottom.

59

5. Tape one set of strips numbered 1 through 4 to the stick, making sure they don't touch each other (B).

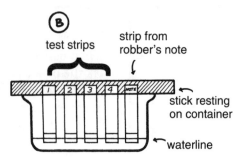

6. Fill your container with ½" of water. Put the four numbered pens into the pencil holder and have the third piece of filter paper ready.

7. Plug in the hair dryer. Get four volunteers and hand out a numbered pen to each one of them.

8. Turning your back to them, ask one volunteer to secretly use his or her black pen to write the following note on the third piece of filter paper: *"Hand over all your jewels! Don't scream! And don't call the police!"* Tell the person to leave a ¾" margin on each side of the paper, and use the same pen to draw a line across the paper about ¾" up from the bottom of the paper (C).

9. Ask the volunteers to put the note into your hand. Then tell them you will discover who the robber is!

10. Cut two ¾" strips from the sides of the note. Write "NOTE" on the top of each strip (B).

11. Tape one of these strips to the stick with the other strips, making sure this strip doesn't touch the others.

12. Rest the stick across the top of the container so that the strips are just touching the water. The ink lines should be above the water.

13. After you let it sit for 10 minutes, check the strips. The ink on some strips should have separated into different colors. Carefully take them off the stick, lay them flat on paper towels, and dry them with the hair dryer.

14. Write down the colors that appear on each strip. Which one has the same pattern as on the note the bank robber wrote? Which pen was it? *NOTE: Are there some strips on which the ink did not separate? Scientists use other chemicals with which to test these. You will use the alcohol. Follow steps 15 through 17 only if the ink on one or more of your strips did not separate. Otherwise, skip to step 18.*

15. Tape the second set of strips in the same order around the outside of the tin can, making sure they do not touch each other and that their bottoms align with the bottom of the can.

16. Pour ½" of isopropyl alcohol into the jar, then put the can with the test strips in the jar.

17. After the 10 minutes are up and the strips are dry, write down the colors that appear. Compare them to your robbery note test strips. Which one has the same pattern as the note on which the bank robber wrote? Which pen was it?

18. Have your volunteers hold up their pens so that you can see the numbers. Tell the person whose number matches, "You're the robber!"

Analysis

The ink in each pen is actually a mixture of different substances (chemical compounds), each of which may be very different or slightly different in color. Each of the different chemicals is attracted to water, alcohol, or the filter paper to a different degree. How strongly one chemical is attracted to (and bonds with) another chemical is called its *affinity*. When the paper strips are dipped into water, the water slowly creeps up tiny spaces in the filter paper due to capillary action. As the water passes through the ink line, chemicals in the ink that have a strong affinity for the water get carried with the upward-moving water. Chemicals with a smaller affinity for water (or a larger affinity for the paper) either get left behind or travel upward more slowly. In this way, the different chemicals in the ink get separated from each other. Inks made up of the same chemicals will show the same pattern of separations. If the ink chemicals have little or no affinity for water, another liquid such as the alcohol can be tried. Because the separated chemicals are usually different colors, the technique is called *chromatography* (in Greek, *chromo* means "color").

This same idea is used to separate chemicals for producing medicines, and even for separating colors found in leaves (see Falling for Color, on page 38).

And They're Off!

Why is it that things that are in motion—such as a ball tossed into the air or a car driving on a highway—eventually come to a stop? Why don't they just continue moving forever? The answer is friction. In this project, you can demonstrate how different surfaces create different amounts of friction.

Materials

- marking pen
- piece of plywood, 2′ × 3′
- glue
- small paper flags of different colors
- enough books to make two stacks 10″ high

- 2′-long strips of rug, linoleum, wax paper, aluminum foil, Plexiglas™, cloth, fine sandpaper, and rough sandpaper (only five items needed)
- five toy cars (Matchbox™ or Hot Wheels™ size)
- five volunteers
- yardstick
- tape measure

Procedure

1. With a marking pen, draw four vertical lines on the plywood to create five lanes. The lanes should each be 7⅕″ wide (A).

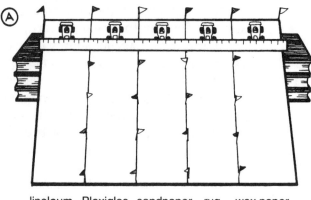

linoleum Plexiglas sandpaper rug wax paper

2. Glue some flags along the lane lines and across the back of the board. Now lean the plywood against two 5″-high stacks of books to make a ramp (A). Place the strips of different materials (any five you choose) under the ramp, one at the base of each lane. Make sure the strips stick out at least 20″ from the ramp.

3. Ask five people to volunteer to be race car drivers. Give each person a car and a lane, and have each predict which car will roll the farthest on the different surfaces.

4. When everyone is ready, hold the yardstick in place as shown so that everyone can place down their cars (A).

5. Now lift up the yardstick and let 'em roll! Which car won? Use the tape measure to record how far each car went on the different surfaces.

6. Use the remainder of the books to raise both ramps another 5", then repeat step 4. Did the increased height help any of the cars travel farther?

Analysis

No surface is truly smooth, not even glass. If you had a very powerful microscope, it would show tiny hills, valleys, and cracks on even the most polished and shiny of surfaces. When two surfaces are rubbed together, these hills and valleys get caught on each other, and it requires energy to break them apart. The larger the hills on the surfaces, the more they stick together and the rougher the material feels to the touch.

The resistance you feel when rubbing two surfaces across each other is called *friction*. In your project, as the cars roll down the ramp, their tires must constantly break free from surface bumps. That breaking free requires energy, so the cars slow down. The rougher the surface, the harder to break free and the slower the car rolls. Without a constant supply of new energy (stepping on the gas), and if nothing else was in its way, a real car would eventually lose all its motional energy to friction and come to a stop.

What happened when you raised the ramp higher? All the cars rolled farther. That's because, starting higher, they could gain more motional energy as they traveled down the steeper ramp (B).

Can you imagine what would happen to a moving object if there was no friction? In the emptiness of space, where there is no friction, an object, once moving (unless it runs into something else), will glide along until the end of time!

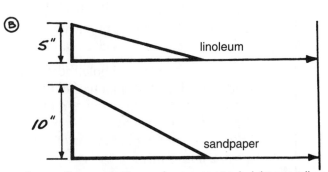

A car rolling on sandpaper from a greater height can roll as far as a car rolling on linoleum from a lower height.

Freefall!

PARENTAL SUPERVISION RECOMMENDED
When you push the "down" button in an elevator at the top of a tall building, you feel yourself falling. But after a split second, the sensation is gone and you feel as though you're just standing there. Why is that? Let's find out.

Materials

- Plexiglas™ tube, with lid, at least 20″ long and 8″ wide
- plastic astronaut (or other human figure), 4″ to 5″ tall
- piece of string, 2′ long
- red paint and paintbrush (optional)
- ladder
- large pillow

Procedure

1. Make a small hole in the end of the tube (the store at which you buy the Plexiglas likely will be able to drill one for you).

2. Tie the astronaut to one end of the string. Pull the other end through the hole in the Plexiglas and tie a knot in it as shown. Make sure the knot is large enough so that the string doesn't fall back inside the tube (A).

3. Paint the word "NASA" vertically on the tube using red paint (optional).

4. Put the tube on a table, stand the ladder next to the table, and put the pillow on the floor in front of the ladder.

5. Pull the string through the lid so that the astronaut hits the lid.

6. Let the string go. The astronaut falls to the bottom of the tube. You have just demonstrated the force of gravity pulling the astronaut back to earth.

7. Now ask an assistant to hold the ladder while you climb it, bringing the tube with you. *NOTE: You may use a taller ladder than the one pictured here. If you do, be sure to have an adult hold it while you climb.*

Ⓐ

20″

8″

8. Pull the string again so that the astronaut is up against the lid.

9. Drop the tube and astronaut together onto the pillow. The tube will hit the floor first, *then* the astronaut (B). Freefall!

Analysis

To understand how you can float in a spaceship (and why the astronaut in your project floated in air), conduct this experiment using only your imagination. Picture yourself in an imaginary elevator 10,000 miles high. You're jumping up and down going crazy, waiting to reach the 5 millionth floor. Just as you're jumping around, the elevator cable snaps and you and the elevator begin to fall to the basement 10,000 miles below. But you don't feel the fall! Why? Since everything falls in the earth's gravity at the same speed, the elevator floor will fall away *from you* as fast as you are falling *toward it*. For every single foot you fall, the elevator floor also falls.

Now suppose the *basement* were also falling away. You'd never hit bottom and never touch the floor. You'd float forever! But this is just a dream—could it ever really be so? Well, if you were an astronaut in orbit around the earth, you and your spaceship would fall together toward the earth (just as the elevator fell toward the basement). If you were traveling fast enough, the earth's surface would curve away from you as fast as you were falling toward it. So "forever falling" is not crazy or imaginary—it's simply an orbit around the earth!

Shape Up!

Why does water take on the particular shapes it does? For instance, why are raindrops round instead of square? It turns out that there are special forces existing at the surface of the water (or any liquid) that give it a particular shape. These forces result in *surface tension*. With a little creative shape making, you can show surface tension at work.

Materials

- dishpan
- 10 cups of water
- ½ cup of dishwashing liquid
- ½ cup of corn syrup (or granulated sugar)
- toothpicks
- glue
- modeling clay
- straws
- yarn (or twist ties)
- pipe cleaners

Procedure

1. In the dishpan, mix together the water, dishwashing liquid, and corn syrup. Make sure this mixture has no bubbles on the surface.

2. Create different shapes out of the toothpicks by dipping the ends of the toothpicks into glue, then joining them together with pieces of modeling clay (A).

3. Cut pieces from the straws and join them together with yarn (B).

4. Put pipe cleaners into straws to make bubble holders. Create any shapes you like, even three-dimensional ones (C). Use your creativity to come up with some interesting and unusual shapes besides the ones shown here.

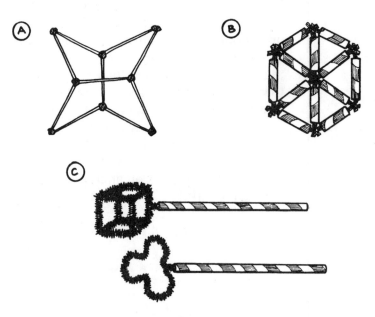

5. Dip the toothpick and straw shapes into soapy water. How many bubble surfaces form in the shapes' sections?

6. Dip a bubble holder into the soapy water and blow a bubble. What happens to the shape of the bubble? What happens when you blow a bubble through a different-shaped bubble blower?

Analysis

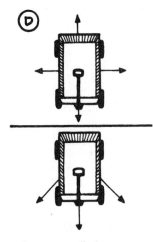

A wagon pulled more in one direction than another will move.

The molecules that make up a liquid, unlike those that make up a solid, can roll and slide around each other, causing the liquid to change its shape. By thinking about pulling on a wagon, you can understand why a blob of liquid molds itself into special shapes. If the wagon is pulled equally from all directions, it doesn't move. Nothing changes. However, if it is pulled more from one side than from another, it will move (D). And it will continue to move until all the forces on it (all the pulls) balance out.

Look at the diagram of the liquid blob (E). A molecule inside the cube-shaped blob gets pulled from all directions equally by the molecules around it, but a molecule on the blob's edge receives a pull only from one side—*into* the blob. As a result, the edge molecules move inward as far as possible, changing the shape of the blob until they can't squeeze inside any farther. This is what happened in your project. When you blew irregularly shaped bubbles, the molecules on the surface of the bubbles wanted to move into as small a shape as possible, so they moved into a sphere! With the shapes you made out of toothpicks and straws, why did just one bubble form in each section? Because of surface tension, the bubbles pulled themselves into the smallest units possible: a single shape in each section.

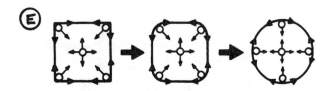

The surface tension on a cube-shaped blob of liquid will pull the liquid into a sphere (the circles are individual molecules).

67

Fighting the Air Force

PARENTAL SUPERVISION RECOMMENDED

By moving a little air around, you can create forces more powerful than the mightiest team of horses!

Materials

- drill
- two identical plastic mixing bowls with wide, flat rims and a plastic ring at the bottom (used to stand up bowls)
- surgical tubing, 4" long
- "sticky" bubble gum
- nylon cording (or shoelaces)
- beeswax
- large spring clip
- two volunteers

Procedure

1. With an adult's help, drill a hole in the side of one of the bowls large enough for the surgical tubing to fit through. Push the tubing in about an inch (A).

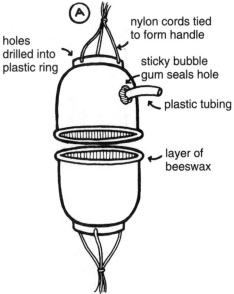

holes drilled into plastic ring

nylon cords tied to form handle

sticky bubble gum seals hole

plastic tubing

layer of beeswax

2. Chew up the sticky bubble gum until it is gooey and can stick to anything, then squish it around the base of the tube to seal any gaps. Depending on the type of bowl, if the gum doesn't seal well, try rubber cement or something similar.

3. Drill three holes, equally spaced, around the plastic ring at the bottom of each bowl.

4. Thread a piece of nylon cording through each hole, knotting the ends so that the cording doesn't pull through the hole. Tie the ends together as shown to create handles (A).

5. Spread a thin layer of beeswax on the flat rims of both bowls, making sure you completely cover the rims.

6. Place the bowls together rim to rim, matching them up precisely. Slowly press them together so that a good seal is made.

7. Now put your mouth over the tube and suck out as much air as possible. With the tube still in your mouth, pinch and bend the tube and then clamp it with the spring clip in order to keep air from getting back in.

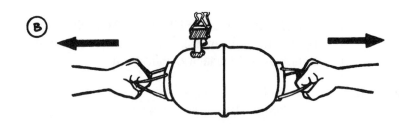

8. Have two volunteers grab opposite handles and attempt to evenly pull the bowls apart (B). If you've done it correctly, they will find it difficult, if not impossible, to separate the bowls. After their attempt, remove the spring clip and see how easily the bowls separate!

Analysis

All around you are molecules of air that are constantly flying around, colliding with each other and everything else. The constant colliding of air molecules with the sides of the bowls creates a pushing force. When you first put the bowls together, there are as many molecules pounding on the inside surface of the bowls (pushing outward) as there are molecules pounding on the outside (pushing inward).

When you sucked out air from the inside, there were fewer air molecules inside to push outward. As a result, the push inward was now much stronger than the push outward (C). The millions of collisions on the large surfaces of the bowls combined to create a huge force that made it nearly impossible to separate the bowls.

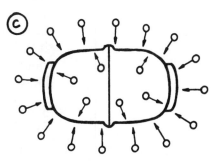

Fewer molecules bombard the walls from the inside of the bowls than from the outside.

A similar experiment was once performed in which two teams of horses attempted to pull apart two large metal hemispheres—and failed!

Here's Looking at You

PARENTAL SUPERVISION RECOMMENDED
Do you think it's possible to see your brain making mistakes? Go face to face with your face and find out!

Materials

- shallow cardboard box (larger than your head)
- wax paper
- plaster of paris
- water
- large mixing bowl
- spatula
- talcum powder and powder puff
- black construction paper (as big as the box)
- scissors
- glue (or tape)
- bright light source

Procedure

1. Line the inside of the box with wax paper.

2. Following the instructions on the plaster of paris box, mix the plaster and water in the bowl with a spatula. Make the plaster just thick enough to hold a shape. You'll need enough plaster to fill the box.

3. Pour the plaster mixture into the box until it's almost full. Smooth the surface flat with the spatula.

4. Now powder your face with a thick layer of talcum powder (to keep the plaster from sticking to your face).

5. With your eyes closed, press your face into the mixture as shown (A). Then carefully pull away from the box. If you look into the box now, you'll see an exact impression of your face. Let the plaster harden overnight.

6. Now make a border for the impression of your face. Take a piece of black construction paper and cut out a large oval in the middle. Glue the paper over the plaster so that the impression is exposed but the remainder of the plaster is covered.

7. Set the box on a table or shelf with the face upright and at eye level (B). Focus the bright light on it so that the face is thoroughly lit.

8. Ask a volunteer to stand facing your plaster face. Ask him or her to look at it as he or she walks to the right. The face will appear to turn and follow your volunteer! Have him or her walk to the left, bob up and down, run right and left. No matter what the volunteer does, the face will seem to follow, turn, bob and nod. Have another viewer try to escape the face's gaze.

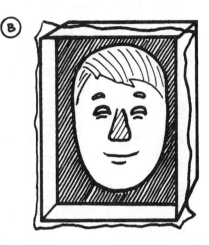

Analysis

How is it the face appeared to be able to respond—to move as the volunteer moved? The key to understanding this effect is to use your nose!

Look at the nose on the plaster face from a distance. Is it sunken in or sticking out? You see it sticking out even though you know it's sunken in. In fact, the whole face looks as if it's sticking out. Even though you *know* the face is a sunken impression, your brain still perceives it as a regular face.

Take a look at Seeing Is Believing (page 46). What you see depends a lot on what patterns your brain is accustomed to seeing. You almost never see an inverted face, so your brain reconstructs the plaster image to match a very familiar pattern—a regular face (on which the nose sticks out).

As you move around a regular face, part of it disappears from view—namely, the side farthest away from you, which is blocked by the side nearest you. With the inverted face, no part of the face sticks out to block your view of it, so you see the entire face as you move. Since the only way you see an entire regular face as you move is if the face turns with you, your brain constructs the familiar pattern of a moving face instead of an inverted one. As a result, the face appears to be following you!

The Big Crash

A huge number of things in the world—from a rocket blasting off, to the inner workings of our sun—can be understood in terms of little "billiard balls" (atoms) colliding into each other. Scientists have invented special ideas to describe just how atoms collide. The most important of these ideas is called *momentum*.

Materials

- Plexiglas™ tube, 2′ long and at least 2¼″ in diameter
- plastic rod, about 8″ long
- eight plastic or wooden balls, 1″ in diameter
- plastic or wooden ball, 2″ in diameter
- different colors of model paint
- paintbrush
- two dowels, 3″ long
- masking tape
- square of acetate, 3″ × 3″
- books or other object about 6″ thick

Procedure

1. Have your local hardware or plastics store cut four holes along one side of your tube as shown. The holes should be large enough for the plastic rod to fit through (A).

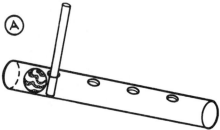

2. To make the balls really stand out, paint stripes or spirals on them, using a different color for each ball.

3. Line up the two dowels on a table, leaving about ¾″ between them. Tape them to the table surface (B).

acetate

4. Put the piece of acetate over one end of the dowels.

5. Rest one end of the tube on the acetate as shown and rest the other end on a stack of books.

6. Put the plastic rod through the top hole in the tube and drop a small ball into the top end. The rod will keep it from rolling.

7. Set four small balls on the dowels as shown (B).

8. Pull out the rod so that the ball rolls down and hits the other balls. Will the balls on the dowels move? Will the ball from the tube stop rolling on impact or keep moving?

9. Repeat step 8 for the other three holes. Notice that the rolling ball does not move as fast, so the balls on the dowels do not move away as quickly. In addition, the balls on the dowels move away a shorter distance each time.

10. Now let the large ball roll from each of the holes. Notice that the large ball hits the balls on the dowels harder than a small ball does and that this force is greatest when the ball is dropped from the topmost hole.

Analysis

When watching balls smashing into each other, what ideas help you predict what will happen? A large, heavy ball released from the top of the tube causes a bigger effect on the target than a small, lighter ball. So is it how heavy the ball is (its mass) that allows you to predict the effect? No, because a small, light ball moving quickly (dropped from the very top hole) can cause as large an effect as a large, heavy ball moving slowly (dropped from the bottom hole). It takes *two* ideas—mass (how heavy an object is) and speed (how fast it is traveling)—to predict the effect of the collision.

Scientists came up with the idea of multiplying mass and speed and calling this number *momentum*. When you know an object's momentum (*both* its mass and speed), then you can predict the outcome of a "collision course." In your project, the more momentum a rolling ball had, the more motional energy it transferred to the stationary balls. That's why they moved when struck!

Beyond Explanation

PARENTAL SUPERVISION RECOMMENDED

The most important and exciting thing for scientists to discover is something they don't understand. That's because they are then forced to question their old ideas and come up with new and better ways of thinking. In this simple project, you'll see an effect that required a revolution in our ideas about the world.

Materials

- scissors
- polarizing plastic sheet, at least 98% efficient, 7″ × 24″ (available at laboratory supply stores)
- large cardboard sheet
- masking tape
- bright light source
- two volunteers

Procedure

1. Cut out three 7″ by 7″ squares from the polarizing plastic.

polarizing plastic sheets in cardboard frames

2. Frame each plastic square with cardboard. With tape, attach little cardboard feet to the frames so that they can stand upright (A).

3. Set two of the framed plastic sheets on the table, about 5″ apart and parallel to each other.

4. Have one volunteer hold a bright light behind the first polarizer while a second volunteer views through the other polarizer.

5. Have the viewing volunteer grab the frame of the second polarizer in front of him or her and slowly rotate it clockwise (keeping it parallel and close to the first polarizer). The volunteer will see the flashlight behind the other polarizer get brighter and dimmer as he or she rotates the plastic. For one position of the polarizer, no light from the flashlight will get through at all, and the plastic will look black. Have the volunteer hold the second polarizer in this position.

6. Now insert the last filter between the other two. The viewing volunteer will see light shine through! (If not, rotate the middle plastic until he or she can see light.) The two polarizers blocked all the light, but adding a third polarizer allowed the light to shine through again. How is it possible?

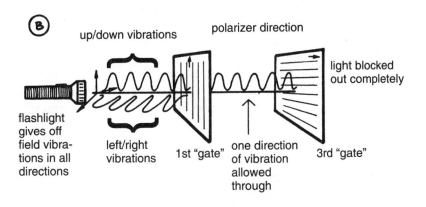

Analysis

When a string vibrates, it can vibrate in many different directions—up and down, left and right, or at some angle in between. Electric and magnetic fields can also vibrate (or fluctuate) in different directions. For example, an electric field fluctuating up and down will move charged particles up and down.

Light is a combination of fluctuating electric and magnetic fields. The light from the flashlight is a complicated mixture of field vibrations in different directions (up/down, left/right, etc.) (B). A polarizer works by acting as a gate that only allows one kind of vibration to pass through. In your setup, the first gate (polarizer) is positioned to allow through up/down vibrations, and the second is rotated to allow through only left/right vibrations. So the up/down vibrations from the flashlight pass through the first filter, but can't pass through the second filter. That's why no light shines through.

Understanding why adding a third filter allows light through requires a completely new way of thinking about nature. This new thinking is called *quantum theory*, and it is stranger than the weirdest science fiction! Without quantum theory, we wouldn't have transistors or lasers.

"X" Marks the Spot

PARENTAL SUPERVISION RECOMMENDED

Have you ever heard the saying "What goes up must come down"? Those words are talking about *gravity*, the force that keeps us from "falling" off the earth. Every object has a *center of gravity* (a point around which the pull of gravity is the same). This project shows you how you can find it.

Materials

- scissors
- several large pieces of cardboard
- hole punch
- coat hanger
- two to three large, heavy books
- weight (a lead fishing kind will do)
- string
- pencil

Procedure

1. Cut the pieces of cardboard into various uneven and jagged shapes.

2. With the hole punch (or the end of a pair of scissors), carefully punch two to four small holes around the edges of each shape (A).

(A) [illustration of various jagged cardboard shapes with holes]

3. Straighten out the hook on the hanger, then place the hanger under some books on a table so that the hook sticks out over the table's edge. The weight of the books will keep the hanger from falling (B).

4. Attach the weight to one end of the string and tie a loop in the other end.

5. Ask a volunteer to hang a shape on the wire.

6. Then hang the looped end of the string on the wire in front of the shape. The weight will keep the string straight (B).

(B) [illustration of books on table edge holding hanger with hanging shape and weighted string]

7. With a pencil, draw a line along the weighted string as shown.

8. Now hang the card by each of the other holes, each time drawing a line along the string. What do you notice? The lines all intersect at the same point—the center of gravity.

9. Show your viewer another shape and have him or her guess where the center of gravity might be. Then repeat steps 6 through 8 and see who's right.

Analysis

For an object to balance on a single point, gravity must pull equally everywhere around this point. Since gravity pulls harder the more material there is, the balance point must have as much material (mass) on one side of it as on the other side. For example, on a long, thin rod, each point along the rod has material to the left and right of it. What point has as much "rod material" to the right as to the left? The middle point, of course. But with your

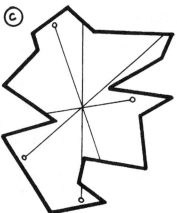

The center of gravity is where lines intersect.

weird shapes, each point on the shape has material to the left and right and above and below it. To figure out just how much material there is left and right, and above and below, you need two pieces of information. The string with the hanging weight shows the direction of the earth's pull. When you hang the shape from the first hole, it will rotate until there are equal amounts of material to the left *and* right of the string. Hanging the shape from a second hole separates material equally in the up/down direction. Where the lines intersect, there is the same amount of mass all around the intersection point (C). That's the center of gravity.

Now think about an irregular cube shape. How would you figure out its center of gravity? You would need at least three lines to separate the material equally—an up/down line, a left/right line, and a front/back line.

'Round the Bend

PARENTAL SUPERVISION RECOMMENDED

If you stand a pencil in a glass of water, then look at it from the side, the pencil looks strange. The part in the water seems to have moved *next* to the part out of water! Actually, it's not the pencil that's bent, but the light reflecting off of it. We call bending light *refraction*. In this project, you'll see refraction at work.

Materials

- scissors
- ruler
- cardboard, 1′ × 1′
- two pieces of wood, 2″ × 2″ × 6″
- books
- sturdy cardboard box
- duct tape
- flashlight
- mirror
- deep glass dish
- water
- eyedropper
- milk
- talcum powder and puff

Procedure

1. About one-third of the way down the piece of cardboard, cut a slit 6″ long and ½″ high (A).

2. Prop up the cardboard on end between the pieces of wood.

3. Put some books behind one flap on top of the cardboard box to create an angled surface as shown (B).

4. Tape the flashlight to the top of the box with the light pointed downward. Make sure the light shines through the cardboard slit.

5. Put the mirror in the bottom of the dish, then fill the dish with water.

Ⓐ slit in cardboard

block of wood

Ⓑ

The light beam refracts, reflects, then refracts again.

tape holds flashlight in place

mirror ↗

cardboard stands between blocks of wood

6. Turn on the flashlight, then put a couple of drops of milk into the water. Ask your viewers if they can see where the light beam is going.

7. Now pat some powder in the air along the light beam's path (B). This will make the beam show up, and your viewers will be able to see it bending.

Analysis

The best way to understand why the light beam bends as it enters and exits the water is to perform this simple experiment. Get a rolling pin (or something similar) and set it rolling across a table. But push one end of it faster than the other. Notice how the rolling pin turns. Only when both ends are pushed at the same speed and with the same force does the rolling pin travel straight.

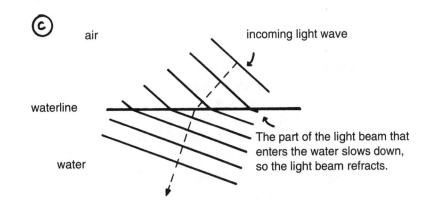

The same thing is happening to your light beam. Light travels slower in water than it does in air. When the beam hits the water at an angle, the side entering the water slows down, and the light beam "bends," or refracts (C). Once the beam is completely inside the water, all parts of it travel at the same speed. Then, as the light exits the water (after reflecting off the mirror), it immediately speeds up again now that it's back in the air. That speeding up refracts the light beam once more.

41 Magnetic Levitation

PARENTAL SUPERVISION RECOMMENDED

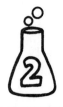

Two famous scientists, Michael Faraday and James Clerk Maxwell, discovered that electricity and magnetism were not completely different forces, but actually closely related effects. In this simple project, you can explore this amazing connection.

Materials

- utility knife
- thin white posterboard
- a strong, doughnut-shaped magnet, about 2″ in diameter and ¼″ thick (a neodymium magnet works well)
- rubber cement
- disk of Styrofoam™, cut to the exact diameter and thickness of the magnet
- two Styrofoam strips, 1″ × 10″ and at least ¼″ thicker than the magnet
- two aluminum sheets, about 8″ × 11″ and at least ¼″ thick (available at metal shops or hardware stores)

Procedure

1. Cut four circles out of the posterboard that are the same diameter as the magnet. Glue the circles to both sides of the magnet and the Styrofoam disk. Then cut and glue a strip of posterboard to cover the sides of the magnet and Styrofoam (A). These two disks should now look the same.

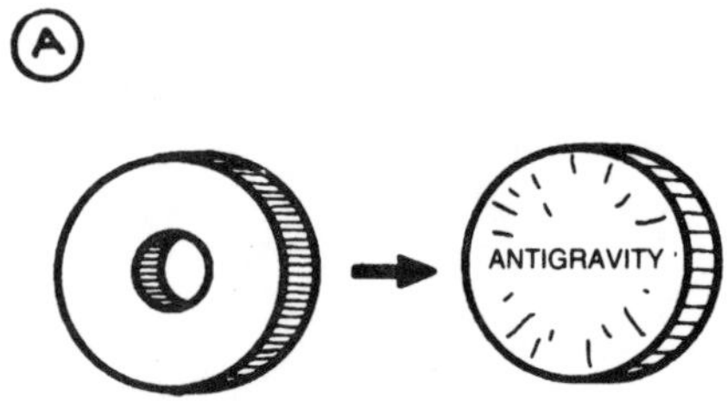

Cover magnet (Styrofoam disk, also) with posterboard.

2. Write "ANTIGRAVITY" on the disk with the magnet inside. Write "GRAVITY" on the other disk.
3. Now glue the two Styrofoam strips to the edges of one aluminum sheet (B). Glue the second aluminum sheet on top and let dry.
4. Feel the weight of each disk. While holding the aluminum sheets vertically about a foot above a table, have a volunteer drop both disks between the aluminum sheets at the

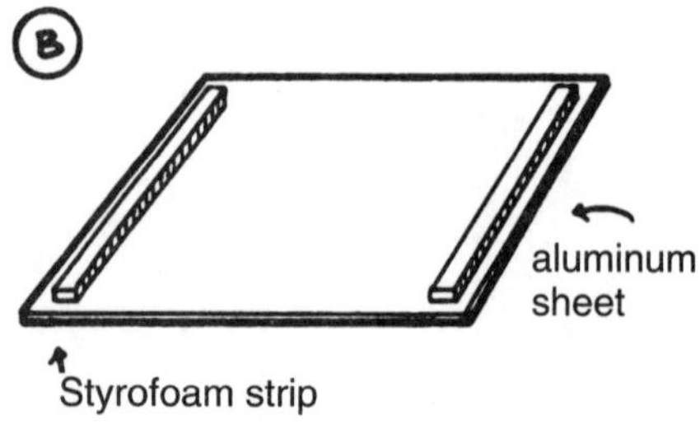

same time (C). The GRAVITY disk will fall through quickly, while the ANTIGRAVITY disk will fall very slowly.

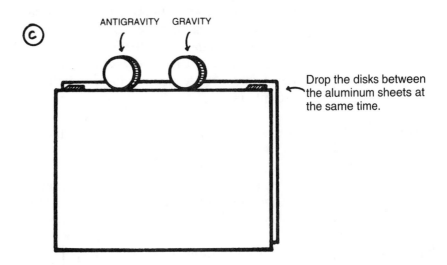

Drop the disks between the aluminum sheets at the same time.

Analysis

A charged particle creates an invisible electric force field around it. That force field attracts or repels other charged particles around it. When you move a charged particle, the field moves with it and creates a magnetic field. The reverse happens, too! When you move a magnet, its magnetic field also moves and in so doing creates an electric field (just like the one created by a charged particle).

In your project, as the magnetic disk falls, it creates an electric field. The field causes charged particles in the aluminum to move. These moving charges in turn create magnetic fields that repel the falling magnet, slowing its fall.

Electricity and magnetism are not completely different effects. They are closely related. One can create the other, so are they different at all?

Half-Life Dating

When scientists discover fossilized bones (like those of a saber-toothed cat), how do they know how old the remains are? One way is by *half-life dating*. Some radioactive elements, such as carbon 14, decay over a period of time. A half-life is the time it takes half the atoms in a chunk of radioactive material to decay into new atoms. With this project, you can demonstrate just how half-life dating is an important tool for scientists.

Materials

- paint and paintbrush
- 100 poker chips (or coins), all the same color
- empty box
- large sheet of white construction paper (or posterboard)
- ruler
- two marking pens, one black and one a color that will show up against white
- sheet of acetate, the same size as the construction paper

Procedure

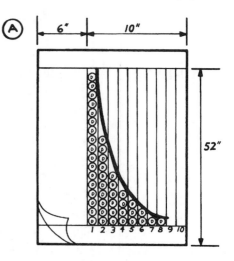

1. Start by painting "D" (for "Decayed") on one side of each of the poker chips and "ND" (for "Not Decayed") on the other side.

2. When the chips are dry, put them into the box.

3. Using the large sheet of construction paper and the colored pen, make a chart similar to the one shown (A).

4. Determine the *result of a set of trials* in the following way. Toss the chips onto a table (B). Then place the chips that come out "D" in the first column on your chart. Put the remaining chips back into the box, toss them again, then place the "D" chips in the second column. Repeat the tossing until all the chips are gone or you have run ten trials. *NOTE: If no chips turn up "D" for a particular trial, leave the column for that trial empty.*

5. The result of this set of trials is the curve the "D" chips make. Lay the acetate over the chart, then draw this curve above the chips using the black pen (A).

6. Set the sheet of acetate aside and put all the chips back into the box.

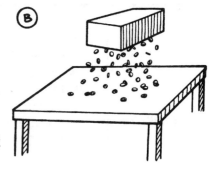

7. Now repeat the set of trials you performed in step 4. When you're done with the tossings, place the acetate down on the chips. The curve you drew earlier will almost perfectly fit the new set of trials!

Analysis

When you flip a penny, although you have no idea whether it will come up heads or tails, over many flips you can predict the penny will land heads up or tails up about the same number of times. There are two reasons why you can predict something about a large number of flips but nothing about a single flip. The first is that you know there are only two possible outcomes (heads, tails). The second is that which outcome you get is completely random—neither result is more likely. If many more heads showed up in a few tossings of a coin, you'd suspect that something was causing heads to come up more often. If the coin were magnetic, or had heads on both sides (in other words, the coin is fake), then the possible outcome is no longer purely random. As a result, you couldn't predict a long-term result anymore. But when the process is completely random (as it is with pennies, and also decaying atoms), then you'll find about an equal number of both possible outcomes. This idea allows you to predict what will happen to a large number of flipped pennies, decaying atoms, and many more things.

The curve you drew on the acetate was a prediction of what would happen. That same curve could predict what would happen in any random system that has only two possible outcomes. Because scientists know how fast atoms decay, they can tell how old something is by counting how many atoms have decayed. In the same way, you can tell from your curve how many trials have passed by counting the number of "D" chips in *any* given column. Did you notice that with every dumping of the box of chips, about half of them came up "D"? You can see why, when dealing with decaying atoms, the process is called half-life dating.

Wayward Compasses

PARENTAL SUPERVISION RECOMMENDED

If you've ever used a compass before, you've probably noticed that the needle inside always points north. Is it possible to make the needle point in another direction? This project will show that it is.

Materials

- utility knife
- large cardboard box
- glue
- two wedge-shaped pieces of wood
- two rulers
- wire cutters
- wire coat hanger
- two electric lead wires with alligator clips at one end
- six small compasses about 1' in diameter
- sheet of cardboard, about 8" × 8"
- battery

Procedure

1. Start by cutting off the top and one side of the box. Then cut two holes in the left side of the box as shown (A).

2. Glue the wood wedges, flat side up, to the back of the box (the same height from the top) and let dry. Then glue the rulers to the top of the wedges.

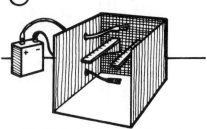

(A) Rulers sit on wedges of wood.

Electric lead wires feed through holes in box.

3. Snip off a 1'-long piece of the hanger. If the hanger is varnished, scrape off 1" of the coating from either end.

4. Guide the lead wires through the holes in the cardboard wall, then connect the alligator clips to either end of the hanger wire (B).

5. Circle the compasses around the wire on the cardboard sheet. Make

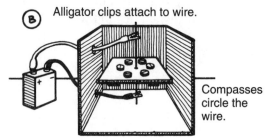

(B) Alligator clips attach to wire.

Compasses circle the wire.

sure all the compass needles point north.

6. Next connect the free ends of the lead wires to each battery terminal. Current will now flow through the hanger wire, producing a magnetic field. What happens to the compass needles? They will all point in the direction of the magnetic field!

7. Disconnect the alligator clips. Now switch the lead wires on the battery terminals. Reconnect the alligator clips. What happens to the compass needles now?

8. Now take one of the compasses (one that is not already pointing in a northerly direction) and walk away from your setup. What happens to the compass needle? At a certain point, the magnetic field will be too weak to affect the compass and the needle will flip to the north.

Analysis

Why did the compasses stop pointing north when the wires were hooked to the battery? What changed? An electric current was now flowing through the wires. Electric current is the flow of charged particles. These particles create an invisible force field around them. When charged particles move, they drag this force field with them, and the changing electrical field is a magnetic field. The battery in your project causes charged particles to flow through the hanger wire, creating a magnetic field around it (C). The compass needles also contain circulating charged particles. These particles create tiny magnets, which are attracted to the wire's magnetic field, causing the needles to rotate. When you reverse the current in the wire (by switching the wires on the battery terminals), the direction of the magnetic field changes, so the needles change direction, too.

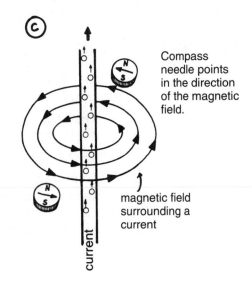

Compass needle points in the direction of the magnetic field.

magnetic field surrounding a current

current

When a magnetic field is very weak, as when you walk away from your setup holding the compass, why does the compass needle point north again? Deep in the earth's molten interior, charged particles circulate around the earth's core, creating a huge magnetic field that spans the globe. This magnetic field causes all compasses to point north.

Spare Me the Details

PARENTAL SUPERVISION RECOMMENDED

Most of what happens around us is the result of many complicated interactions of which we cannot keep track. Fortunately, we can often ignore most of the details and look at the "big picture." For example, while we can't possibly know the detailed motion of a single water molecule in a river, we can still predict something about the motion of the river as a whole. In this project, you'll create a device that is a good model for many systems in nature.

Materials

- piece of wood, 17″ × 16″ and ½″ thick
- 160 wooden pegs, ½″ long and ¼″ in diameter
- wood glue
- rigid Plexiglas™ sheet, 17″ × 16″ and ⅛″ thick
- 14 rectangular wood pieces, ½″ × 8″ and ⅛″ thick
- two pieces of wood, 17″ × ½″ and ⅛″ thick
- one piece of Plexiglas, 16″ × ½″ and ⅛″ thick
- 200 small marbles, ¼″ in diameter (round candies or small wooden beads work, also)
- funnel large enough for the marbles to easily fall through
- marker that can be erased from plastic

Procedure

1. Take the large piece of wood, and start by making a pegboard, 8 rows deep and 20 rows across (A). From left to right, the pegs should be spaced ¾″ apart (from the center of one peg to the next). From top to bottom, the rows should be 1″ apart (again from the center of one peg to the next)(B). Shift every other row to the right ⅜″, so that each peg is *between* the two above it. Glue the wooden pegs in place and let dry.

2. Glue down the 14 rectangular wooden "slot" pieces on their edges as shown, 1″ apart.

3. Glue on the top Plexiglas piece and the two thin pieces of wood on the sides to complete the box as shown. Allow the entire assembly to dry for several hours.

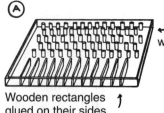

Ⓐ

wooden pegs

(illustration not to scale)

Wooden rectangles glued on their sides form bins for marbles to fill.

4. Stand your box vertically, with the open end up. While a volunteer holds the funnel, pour the 200 marbles through the funnel into the center of the box (B). Now the fun begins!

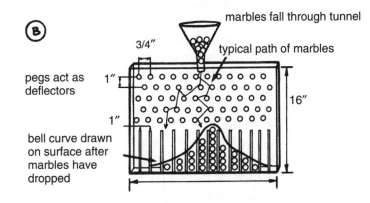

5. The marbles will collide with the pegs on the way down and get knocked in all different directions. Eventually, all the marbles will fall into one of the slots below.

6. With the marker, draw a curve on the plastic along the tops of the stacked marbles. Notice how the curve is shaped like a bell—high in the middle and dropping off toward the edges.

7. Pour the balls out and repeat step 4. When the marbles are done falling, they should form the same shape as before.

Analysis

As you drop the marbles, it's impossible to know what any given marble will do. There are too many collisions with pegs and other marbles. But the marbles fall into the same curve every time. Why?

The curve tells you how likely it is that a marble will land in a certain column. The marbles were dropped into the center, so it's more likely that the marbles will end up there. But occasionally, a number of unlikely collisions will send a marble to the edge. The farther from the center you go, the less likely it is that a marble will end up there. That's why fewer balls landed there.

This same curve often describes how test scores are distributed (how many people get what grade). Most students get about the same grade (the bump in the center of the curve). Fewer people get extremely high or extremely low grades (the flattened edges of the curve). Using a type of math called *statistics*, scientists can predict outcomes. While they are not able to predict a particular outcome (such as exactly how many students will get a B), they can predict how likely an outcome is (for instance, that fewer students will probably get a D than a C).

Cosmic Ray Detector

PARENTAL SUPERVISION RECOMMENDED

Every second of every day, invisible high-energy particles shoot through and past you. Where do these particles come from? Most are from the disintegration of radioactive substances, such as uranium, here on earth. Others are cosmic rays—particles released into space, perhaps millions of years ago, by exploding stars! By building a cloud chamber, you can show traces of these ghostly particles as they travel through our world.

Materials

- scissors
- black felt
- large glass jar with a metal screw-on lid
- thin sponge
- black construction paper
- rubber cement
- isopropyl alcohol
- block of dry ice, $10'' \times 10'' \times 2''$
- plastic insulated gloves
- sheet of cardboard, $1' \times 1'$
- towel (large enough to wrap around the dry ice)
- strong light source (as from a slide or movie projector)
- a room in which you can turn off the lights
- watch with second hand

Procedure

1. First, cut and glue a circular piece of felt large enough to fit inside the metal lid, as well as a circular piece of sponge large enough to cover the bottom of the jar (A). Let dry.

2. Cut a strip of black paper wide enough to loosely wrap around the jar and a little taller than the jar is high.

    (A) black paper cylinder covers jar — viewing slot — sponge disk glued to bottom of glass jar — illumination slot — felt glued to inside of metal lid

3. Cut two slots in the paper cylinder 1″ high by 3″ wide where shown (A) (the upper one is a viewport, the lower one is for your light source). Glue the strip closed.

4. Slowly pour the alcohol into the jar and soak the sponge and the felt. Tightly screw on the lid.

5. Put the dry ice on the cardboard and put the jar upside down on top of the ice, making sure the metal lid comes in contact with the ice (B). *WARNING! Dry ice is extremely cold! Wear plastic insulated gloves while handling it. Do NOT touch it with your bare hands.*

6. Wrap the towel around the remaining ice.

7. Slip the black paper cylinder over the jar and arrange the projector light so that it shines through the

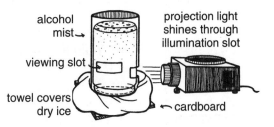

illumination slot as shown. Turn off the room lights.

8. Ask a volunteer to look into the viewing slot toward where the light enters the jar. He or she will see a miniature rainfall, with tiny threadlike vapor trails appearing out of nowhere. These are the tracks left by high-energy particles as they fly through the jar! Let other spectators peer into the jar to see the particle tracks. Using the watch, count the number of tracks they see in a minute.

Analysis

If conditions are just right, a gaseous vapor (such as water mist) will condense to form liquid droplets. That's what happens in the earth's atmosphere when water molecules condense onto dust particles, form droplets, and fall to the earth as rain.

In your chamber, you created almost all of the conditions necessary to make "alcohol rain." As alcohol from the sponge falls to the bottom, it cools and forms a dense, cold gas. All that was needed to form alcohol droplets was something for the alcohol molecules to condense onto (like the dust in the atmosphere that water condenses onto). When an invisible, high-energy particle shoots through the alcohol gas, it collides with alcohol molecules in its path, knocking off some of their electrons (negatively charged particles). Lacking electrons, these alcohol molecules now are positively charged, so they attract the neutral alcohol molecules around them (C). As the alcohol molecules gather, they form small droplets. The trail of droplets is like a trail of footprints marking the path of the invisible particles.

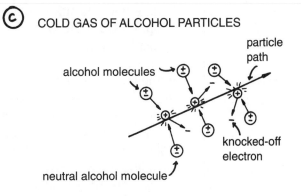

© COLD GAS OF ALCOHOL PARTICLES

46 ◇ Helloooooo!

When two people are standing apart, they can't hear each other very well. Could they hear each other better if an object was placed between them? Show that *focusing* sound can make a whisper sound like a shout!

Materials

- large balloon, 3' to 4' in diameter
- carbon dioxide (CO_2) tank (look in your phone book for a supplier of compressed gases)
- round washtub, 2' in diameter
- volunteer
- high-pitched buzzer

Procedure

1. Fill the balloon with CO_2 gas and set it on the washtub as shown (A). Ask a volunteer to sit on one side of the balloon while you sit on the opposite side. Turn on the buzzer.

2. Ask the volunteer to move around on his or her side of the balloon until he or she finds the place where he or she hears the buzzer the loudest.

3. Then remove the balloon and turn on the buzzer again. Ask the volunteer to say how loud the sound is now.

4. Replace the balloon and buzz the buzzer again, only this time hold the buzzer higher up. Where is the loudest point now?

The buzzer produces sound vibrations, and the carbon dioxide causes them to converge.

Analysis

Sound vibrations are much like ripples expanding on a pond. The buzzer in your project sends out ripples of sounds (sound waves). The ripples spread over a larger and larger area. As they do, they get weaker and weaker and finally die out (and can no longer be heard). The CO_2 molecules in the balloon cause a section of the vibrations to stop spreading out and instead come together (converge) at a point beyond the balloon (B). The increased vibrations at this point create a louder sound. Thus, the CO_2 gas in the balloon acts as a "lens" to focus sound!

Bubbles, Bubbles Everywhere

PARENTAL SUPERVISION RECOMMENDED
How are soap bubbles similar to the cells in our bodies? Do this project and find out.

Materials

- Plexiglas™ box, about 1′ wide, 1′ deep, and 3′ tall (an emptied aquarium will work well, also)
- ⅔ cup of dishwashing liquid
- 1 tablespoon of glycerine (available at drugstores)
- 1 gallon of water
- large mixing bowl (or bucket)
- wire
- dishcloth
- dry ice
- plastic insulated gloves

Procedure

1. At a plastics or hardware store, have some sheets of Plexiglas cut and glued together into a tall box.

2. Make a bubble-blowing solution by mixing together in the bowl the dishwashing liquid, glycerine, and water. Let the mixture sit one day.

3. With the wire, make a 2″ bubble blower.

4. Just before your demonstration, put the box on the floor next to a table, place the dishcloth in the bottom of the box, and put the dry ice on top of the cloth. *WARNING! Dry ice is extremely cold! Wear plastic insulated gloves while handling it. Do NOT touch it with your bare hands.*

5. Dip the bubble blower into the mixture, then blow bubbles into the box so that they sink to the bottom (A). What happens?

Analysis

Bubbles have what's called a *semipermeable membrane*. That means that their thin outer surface allows certain molecules to enter. One of these molecules is carbon dioxide (CO_2). As the bubbles sink into the CO_2 gas created by the dry ice, CO_2 molecules gradually diffuse into (enter) the bubbles (B). Air molecules, on the other hand, are not able to diffuse into or out of the bubbles. With no air leaking out of the bubbles, and with CO_2 leaking in, the bubbles grow in size.

The cells in your body also have a thin, permeable membrane surrounding them that allows only certain molecules (such as oxygen) in or out. So at least one aspect of cells can be understood by blowing bubbles!

A *Cube-Shaped* Globe?

PARENTAL SUPERVISION RECOMMENDED

If the earth were cube-shaped instead of round, how would life be different? With this project, you can see!

Materials

- scissors or sharp knife
- soft, hollow rubber ball, about 6″ in diameter
- empty half-gallon milk carton
- two thumbtacks
- two dowels (or unsharpened pencils)

- modeling clay
- small squeeze bottle
- water
- blue food coloring
- plastic tablecloth

Procedure

1. With a parent's help, cut the ball in half to create a hemisphere and cut a cube shape from the bottom of the milk carton.

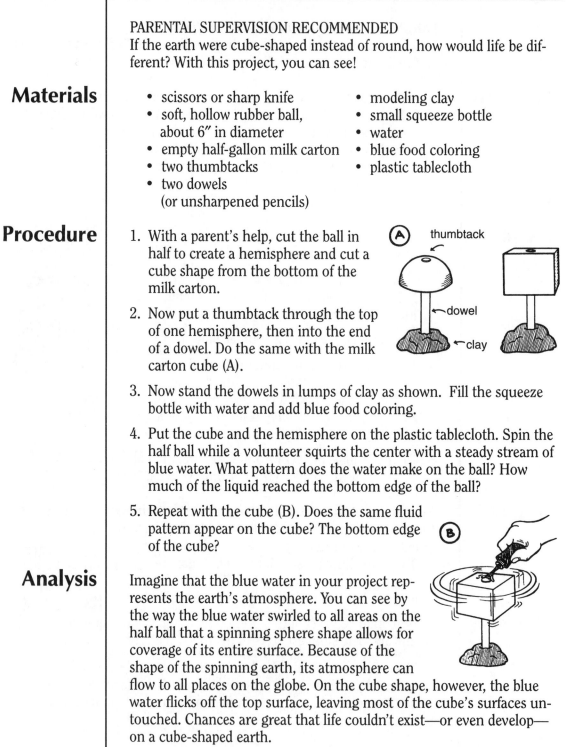

2. Now put a thumbtack through the top of one hemisphere, then into the end of a dowel. Do the same with the milk carton cube (A).

3. Now stand the dowels in lumps of clay as shown. Fill the squeeze bottle with water and add blue food coloring.

4. Put the cube and the hemisphere on the plastic tablecloth. Spin the half ball while a volunteer squirts the center with a steady stream of blue water. What pattern does the water make on the ball? How much of the liquid reached the bottom edge of the ball?

5. Repeat with the cube (B). Does the same fluid pattern appear on the cube? The bottom edge of the cube?

Analysis

Imagine that the blue water in your project represents the earth's atmosphere. You can see by the way the blue water swirled to all areas on the half ball that a spinning sphere shape allows for coverage of its entire surface. Because of the shape of the spinning earth, its atmosphere can flow to all places on the globe. On the cube shape, however, the blue water flicks off the top surface, leaving most of the cube's surfaces untouched. Chances are great that life couldn't exist—or even develop—on a cube-shaped earth.

Only Your Cabbage Knows

PARENTAL SUPERVISION RECOMMENDED
This project experiments with chemistry. By adding different products to an "indicator" made of cabbage juice, you can turn things green!

Materials

- plastic tablecloth
- two or three cans of cooked red cabbage
- six or seven glasses
- the following products to test: baking soda, vinegar,

ammonia, club soda, cream of tartar, carrot juice, apple juice, Tabasco™ sauce, beet juice, and ground aspirin
- wooden stirrers

Procedure

1. Cover a table with the tablecloth. Pour about ¼ cup of juice from the cans of cabbage into each of the glasses. Display the cans or bottles that your test products come in next to the glasses.

2. Explain to your viewers the difference between acids and bases. To help them understand, tell them that milk is a base and orange juice is an acid. The cabbage juice is an indicator.

3. Ask several volunteers to look at the test products and guess which ones are acids and which are bases. Ask them to write down their guesses.

4. Add a pinch of the baking soda to the first glass of cabbage indicator and stir. The color will change to green (if it doesn't, add more baking soda). The green shows that the mixture is a base.

5. Now add about half a teaspoon of vinegar to the second glass. The mixture will remain red, showing that it is an acid.

6. Repeat with the remaining products and glasses. Which are acids and which are bases? For fun, take the indicator that the baking soda was added to and add some vinegar. What happens to the mixture?

Analysis

A compound consists of two or more kinds of atoms that combine chemically. The simplest atom is hydrogen. Two groups of compounds that are very easy to identify are *acids* and *bases*. Acids readily *give away* hydrogen atoms to other compounds, while bases readily *grab* hydrogen atoms. When you pour an acid into a base (or vice versa), they neutralize each other. Why? Because the base grabs the hydrogen atoms that the acid wants to give away anyway!

It's Not Black or White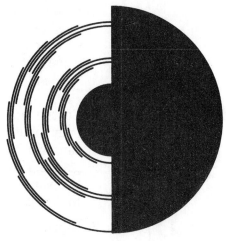

Benham was a toymaker who lived in the 18th century. When he spun the black-and-white design you see below on a top, colors appeared! With this project, you can demonstrate this unusual phenomenon.

Materials

- glue stick
- sanding disk (for the drill)
- variable-speed electric drill
- desk lamp with an incandescent bulb

Procedure

1. Copy the disk pattern shown here by photocopying it. Enlarge it as needed to make it the same size as the sanding disk.

2. Glue the disk pattern onto the sanding disk and attach it to the drill.

3. Show your viewers Benham's disk. Then plug in the drill and the desk lamp and spin the disk under the bright light.

4. Ask your viewers what they see. The black-and-white disk is suddenly colorful! What colors do your viewers see, and in what order are they?

Analysis

The retina in your eye has three types of receptors (areas sensitive to light). Each receptor is sensitive to a different color of light: one blue, one red, and one green. Not only do these receptors respond to different colors, but they also behave differently. For instance, when hit by light, the blue receptors take a longer time to send signals to your brain than the red or green receptors do. And once the blue or green receptors start sending signals, they take longer than the red receptors to stop sending signals.

When a steady white light hits your eyes, the slower receptors have time to catch up. After a short time, all receptors are responding equally, so you see the white. But with Benham's disk, the white quickly flashes on and off as the disk spins. Therefore, your slow receptors can't catch up with your fast ones. The result is that your color receptors never have a chance to respond equally, and instead of seeing white you see colors.